PROBABLY NOT

PROBABLY NOT
Future Prediction Using Probability and Statistical Inference

Lawrence N. Dworsky
Phoenix, AZ

WILEY-INTERSCIENCE

A JOHN WILEY & SONS, INC. PUBLICATION

Published by John Wiley & Sons, Inc., Hoboken, New Jersey
Published simultaneously in Canada

For general information on our other products and services or for technical support, please contact our Customer Care Department within the United States at (800) 762-2974, outside the United States at (317) 572-3993 or fax (317) 572-4002.

Wiley also publishes its books in a variety of electronic formats. Some content that appears in print may not be available in electronic formats. For more information about Wiley products, visit our web site at www.wiley.com.

Library of Congress Cataloging-in-Publication Data:

Dworsky, Lawrence N., 1943–
   Probably not : future prediction using probability and statistical inference / Lawrence N. Dworsky.
      p. cm.
   Includes bibliographical references and index.
   ISBN 978-0-470-18401-1 (pbk.)
   1. Prediction theory. 2. Probabilities. 3. Mathematical statistics. I. Title.
   QA279.2.D96 2008
   519.5′4—dc22

                                                                    2007050155

Printed in the United States of America

10 9 8 7 6 5 4 3 2 1

To my grandchildren, Claudia and Logan, and to grandchildren everywhere—probably the best invention of all time.

CONTENTS

PREFACE

For as long as I can remember, I have been interested in how well we know what we say we know, how we acquire data about things, and how we react to and use this data. I was surprised when I first realized that many people are not only not interested in these things, but are actually averse to learning about them. It wasn't until fairly recently that I concluded that we seem to be genetically programmed, on one hand, to intelligently learn how to acquire data and use it to our advantage but also, on the other hand, to stubbornly refuse to believe what some simple calculations and/or observations tell us.

My first conclusion is supported by our march through history, learning about agriculture and all the various forms of engineering and using this knowledge to make life better and easier. My latter conclusion comes from seeing all the people sitting on stools in front of slot machines at gambling casinos, many of whom are there that day because the astrology page in the newspaper told them that this was "their day."

This is a book about probability and statistics. It's mostly about probability, with just one chapter dedicated to an introduction to the vast field of statistical inference.

There are many excellent books on this topic available today. I find that these books fall into two general categories. One category is textbooks. Textbooks are heavily mathematical with derivations, proofs and problem sets, and an agenda to get you through a term's course work. This is just what you need if you are taking a course.

The other category is books that are meant for a more casual audience—an audience that's interested in the topic but isn't interested enough to take a course. We're told today that people have "mathephobia," and the books that appeal to these people try very hard to talk around the mathematics without actually presenting any of it. Probability and statistics are mathematical topics. A book on these subjects without math is sort of like a book on French grammar without any French words in it. It's not impossible, but it sure is doing things the hard way.

This book tries to split the difference. It's not a textbook. There is, however, some math involved. How much? Some vague reminiscences about introductory high school algebra along with a little patience in learning some new notation should comfortably get you through it. You should know what a fraction is and recognize what I'm doing when I add or multiply fractions or calculate decimal equivalents. Even if you don't remember how to do it yourself, just realizing what I'm doing and accepting that I'm probably doing it right should be enough. You should recognize a square root sign and sort of remember what it means. You don't need to know how to calculate a square root—these days everybody does it on a pocket calculator or a computer spreadsheet anyway. You should be able to read a graph. I review this just in case, but a little prior experience helps a lot. In a few cases some elementary calculus was needed to get from point A to point B. In these cases I try to get us all to point A slowly and clearly and then just say that I needed a magic wand to jump from A to B and that you'll have to trust me.

If you thumb through the book, you'll see a few "fancy" formulas. These are either simply shorthand notations for things like repeated additions, which I discuss in great detail to get you comfortable with them, or in a few cases some formulas that I'm quoting just for completeness but that you don't need to understand if you don't want to.

As I discuss in the first chapter, probability is all about patterns of things such as what happens when I roll a pair of dice a thousand times, or what the life expectancies of the population of the United States looks like, or how a string of traffic lights slows you down in traffic. Just as a course in music with some discussions of rhythm and harmony helps you to "feel" the beauty of the music, a little insight into the mathematics of the patterns of things in our life can help you to feel the beauty of these patterns as well as to plan things that are specifically unpredictable (when will the next bus come along and how long will I have to stand in the rain to meet it?) as best possible.

Most popular science and math books include a lot of biographical information about the people who developed these particular fields. This can often be interesting reading, though quite honestly I'm not sure that knowing how Einstein treated his first wife helps me to understand special relativity.

I have decided not to include biographical information. I often quote a name associated with a particular topic (Gaussian curves, Simpson's Paradox, Poisson distribution) because that's how it's known.

Probabilistic considerations show up in several areas of our lives. Some we get explicitly from nature, such as daily rainfall or distances to the stars. Some we get from human activities, including everything from gambling games to manufacturing tolerances. Some come from nature, but we don't see them until we "look behind the green curtain." This includes properties of gases (e.g., the air around us) and the basic atomic and subatomic nature of matter.

Mathematical analyses wave the banner of *truth*. In a sense, this is deserved. If you do the arithmetic correctly, and the algorithms, or formulas, used are correct, your result is *the* correct answer and that's the end of the story. Consequently, when we are presented with a conclusion to a study that includes a mathematical analysis, we tend to treat the conclusion as if it were the result of summing a column of numbers. We believe it.

Let me present a situation, however, where the mathematics is absolutely correct and the conclusion is absolutely incorrect. The mathematics is simple arithmetic, no more than addition and division. There's no *fancy stuff* such as probability or statistics to obfuscate the thought process or the calculations. I've changed the numbers around a bit for the sake of my example, but the situation is based upon an actual University of California at Berkeley lawsuit.

We have a large organization that is adding two groups, each having 100 people, such as two new programs at a school. I'll call these new programs A and B.

Program A is an attractive program and for some reason is more appealing to women than it is to men. 600 women apply; only 400 men apply. If all the applicants were equally qualified, we would expect to see about 60 women and 40 men accepted to the 100 openings for the program. The women applicants to this program tend to be better qualified than the men applicants, so we end up seeing 75 women and 25 men accepted into program A. If you didn't examine the applications yourself, you might believe that the admissions director was (unfairly) favoring women over men.

Program B is not as attractive a program and only 100 people apply. It is much more attractive to men than it is to women: 75 men and 25 women apply. Since there are 100 openings, they all get accepted.

Some time later, there is an audit of the school's admission policies to see if there is any evidence of unfair practices, be they sexual, racial, ethnic, whatever. Since the new programs were handled together by one admissions director, the auditor looks at the books for the two new programs as a group and sees that:

$600 + 25 = 625$ women applied to the new programs. $75 + 25 = 100$ women were accepted. In other words, $100/625 = 16\%$ of the women applicants were accepted to the new programs.

$400 + 75 = 475$ men applied to the new programs. $25 + 75 = 100$ men were accepted. In other words, $100/475 = 21\%$ of the men applicants were accepted to the new programs.

The auditor then reviews the qualifications of the applicants and sees that the women applicants were in no way inferior to the men applicants; in fact it's the opposite. The only plausible conclusion is that the programs' admissions director favors men over women.

The arithmetic above is straightforward and cannot be questioned. The flaw lies in how details get lost in summarization—in this case, looking only at the totals for the two programs rather than keeping the data separate. I'll show (in Chapter 13) how a probabilistic interpretation of these data can help to calculate a summary correctly.

My point here, having taken the unusual step of actually putting subject matter into a book's preface, is that mathematics is a tool and only a tool. For the conclusion to be correct, the mathematics along the way must be correct, but the converse is not necessarily true.

Probability and statistics deals a lot with examining sets of data and drawing a conclusion—for example, "the average daily temperature in Coyoteville is 75 degrees Fahrenheit." This sounds like a great place to live until you learn that the temperature during the day peaks at 115 degrees while at night it drops to 35 degrees. In some cases we will be adding insight by summarizing a data set, but in some cases we will be losing insight.

My brother-in-law Jonathan sent me the following quote, attributing it to his father. He said that I could use it if I acknowledge my source: Thanks, Jonathan.

"The average of an elephant and a mouse is a cow, but you won't learn much about either elephants or mice by studying cows." I'm not sure exactly what the arithmetic in this calculation would look like, but I think it's a memorable way of making a very good point.

I could write a long treatise on how bad conclusions have been reached because the people who had to draw the conclusions just weren't looking at all the data. Two examples that come to mind are (1) the Dow silicone breast implant lawsuit where a company was put out of business because the plaintiffs "demonstrated" that the data showed a link between the implants and certain serious disease and (2) the crash of the space shuttle Challenger where existing data that the rubber O-rings sealing the liquid hydrogen tanks get brittle below a certain temperature somehow never made it to the table.

The field of probability and statistics has a very bad reputation ("Lies, Damned Lies, and Statistics"[1]). It is so easy to manipulate conclusions by simply omitting some of the data, or to perform the wrong calculations correctly, or to misstate the results—any and all of these possibly innocently—because some problems are very complicated and subtle. I hope the materials to follow show what information is needed to draw a conclusion and what conclusion(s) can and can't be drawn from certain information. Also I'll show how to reasonably expect that sometimes, sometimes even inevitably, as the bumper stickers say, *stuff* happens.

---

[1] This quote is usually attributed to Benjamin Disraeli, but there seems to be some uncertainty here. I guess that, considering the book you're now holding, I should say that "There is a high probability that this quote should be attributed to Benjamin Disraeli."

I spend a lot of time on simple gambling games because, even if you're not a gambler, there's a lot to be learned from the simplest of random events—for example, the result of coin flips.

I've also tried to choose many examples that you don't usually see in probability books. I look at traffic lights, waiting for a bus, life insurance, scheduling appointments, and so on. What I hope to convey is that we live in a world where so many of our daily activities involve random processes and the statistics involved with them.

Finally, I introduce some topics that show how much of our physical world is based on the consequences of random processes. These topics include gas pressure, heat engines, and radioactive decay. These topics are pretty far from things you might actually do such as meeting a friend for lunch or counting birds in the woods. I hope you'll find that reading about them will be interesting.

One last comment: There are dozens and dozens of clever probability problems that make their way around. I've included several of these (the shared birthday, the prize behind one of three doors, etc.) where appropriate and I discuss how to solve them. When first confronted with one of these problems, I inevitably get it wrong. In my own defense, when I get a chance to sit down and work things out carefully, I (usually) get it right. This is a tricky subject. Maybe that's why I find it to be so much fun.

ACKNOWLEDGMENTS

My wife, Suzanna, patiently encouraged me to write this book for the many months it took me to actually get started. She listened to my ideas, read and commented on drafts and perhaps most importantly, suggested the title.

My daughter Gillian found the time to read through many chapter drafts and to comment on them. My son-in-law Aaron and I had many interesting conversations about the nature of randomness, random sequences of numbers, and the like that clarified my thoughts significantly.

My friends Mel Slater and Arthur Block and my brother-in-law Richie Lowe also found the time to read and comment on chapter drafts.

The folks at John Wiley were very encouraging and helpful from the start.

I am grateful to all of you.

CHAPTER 1

AN INTRODUCTION TO PROBABILITY

PREDICTING THE FUTURE

The term Predicting the Future conjures up images of veiled women staring into hazy crystal balls, or bearded men with darting eyes passing their hands over cups of tea leaves, or something else equally humorously mysterious. We call these people Fortune Tellers and relegate their "professions" to the regime of carnival side-show entertainment, along with snake charmers and the like. For party entertainment we bring out a Ouija board; everyone sits around the board in a circle and watches the board extract its mysterious "energy" from our hands while it answers questions about things to come.

On the other hand, we all seem to have firm ideas about the future based on consistent patterns of events that we have observed. We are pretty sure that there will be a tomorrow and that our clocks will all run at the same rate tomorrow as they did today. If we look in the newspaper (or these days, on the Internet), we can find out what time the sun will rise and set tomorrow— and it would be very difficult to find someone willing to place a bet that this information is not accurate. Then again, whether or not you will meet the love of your life tomorrow is not something you expect to see accurately predicted in the newspaper.

We seem willing to classify predictions of future events into categories of the *knowable* and the *unknowable*. The latter category is left to carnival

Probably Not: Future Prediction Using Probability and Statistical Inference,
by Lawrence N. Dworsky.
Copyright © 2008 John Wiley & Sons, Inc.

fortune tellers to illuminate. The former category includes "predictions" of when you'll next need a haircut, how much weight you'll gain if you keep eating so much pizza, and so on. There does seem to be, however, an intermediate area of knowledge of the future. Nobody knows for certain when you're going to die. An insurance company, however, seems able to consult its mystical Actuarial Tables and decide how much to charge you for a life insurance policy. How can it do this if nobody knows when you're going to die? The answer seems to lie in the fact that if you study thousands of people similar in age, health, life style, and so on, to you, you would be able to calculate an average life span—and that if the insurance company sells enough insurance policies with rates based upon this average, in a financial sense this is "as good" as if the insurance company knows exactly when you are going to die. There is, therefore, a way to describe life expectancies in terms of the expected behavior of large groups of people in similar circumstances.

When predicting future events, you often find yourself in situations such as this where you know something about future trends but you do not know exactly what is going to happen. If you flip a coin, you know you'll get either heads or tails but you don't know which. If you flip 100 coins, or equivalently flip one coin 100 times, however, you'd expect to get approximately 50 heads and 50 tails.

If you roll a pair of dice, you know that you'll get some number between two and twelve, but you don't know which number you'll get. However, in the case of the roll of a pair of dice, you do know that, in some sense, it's more likely that you'll get six than that you'll get two.

When you buy a new light bulb, you may see written on the package "estimated lifetime 1500 hours." You know that this light bulb might last 1346 hours, 1211 hours, 1587 hours, 2094 hours, or any other number of hours. If the bulb turns out to last 1434 hours, you won't be surprised; but if it only lasts 100 hours, you'd probably switch to a different brand of light bulbs.

There is a hint that in each of these examples, even though you couldn't accurately predict the future, you could find some kind of pattern that teaches you something about the nature of the future. Finding these patterns, working with them, and learning what knowledge can and cannot be inferred from them is the subject matter of the study of probability and statistics.

I can separate our study into two classes of problems. The first of these classes is understanding the likelihood that something *might* occur. We need a rigorous definition of likelihood so that we can be consistent in our evaluations. With this definition in hand, I can look at problems such as "How likely is it that you can make money in a simple coin flipping game?" or "How likely is it that a certain medicine will do you more good than harm in alleviating some specific ailment?" I'll have to define and discuss *random events* and the patterns that these events fall into, called Probability Distribution Functions (PDFs). This study is the study of Probability.

The second class of problems involves understanding how well you really know something. I will only present quantifiable issues, not "Does she really love me?" and "Is this sculpture truly a work of art?"

The uncertainties in how well we really know something can come from various sources. Let's return to the example of light bulb. Suppose you're the manufacturer of these light bulbs. Due to variations in materials and manufacturing processes, no two light bulb filaments (the thin wires in these bulbs that get white hot and glow brightly) are identical. There are variations in the lifetime of your product that you need to understand. The easiest way to learn the variations in lifetime would be to run all your light bulbs until they burn out and then look at the numbers, but for obvious reasons this is not a good idea. If you could find the pattern by just burning out some (hopefully a small percentage) of the light bulbs, then you have the information you need both to truthfully advertise your product and to work on improving your manufacturing process.

Learning how to do this is the study of Statistics. I will assume that we are dealing with a *stationary random process*. In a stationary random process, if nothing causal changes, we can expect that the nature of the pattern of the data already in hand will be the same as the nature of the pattern of future events of this same situation, and we use *statistical inference* to predict the future. In the practical terms of our light bulb manufacturer example, I am saying that so long as we don't change anything, the factory will turn out bulbs with the same distribution of lifetimes next week as it did last week. This assertion is one of the most important characteristics of animal intelligence, namely the ability to discern and predict based upon patterns. If you think that only people can establish a pattern from historical data and predict the future based upon it, just watch your dog run to the door the next time you pick up his leash.

This light bulb problem also exemplifies another issue that I will have to deal with. We want to know how long the light bulb we're about to buy will last. We know that no two light bulbs are identical. We also realize that our knowledge is limited by the fact that we haven't measured every light bulb made. We must learn to quantify how much of our ignorance comes from each of these factors and develop ways to express both our knowledge and our lack of knowledge.

RULE MAKING

As the human species evolved, we took command of our environment because of our ability to learn. We learn from experience. Learning from experience is the art/science of recognizing patterns and then generalizing these patterns to a *rule*. In other words, the pattern is the relevant raw data that we've collected. A rule is what we create from our analysis of the pattern that we use to predict the future. Part of the rule is either one or several preferred extrapolations and responses. Successful pattern recognition is, for example, seeing that seeds from certain plants, when planted at the right time of the year and given the right amount of water, will yield food; and that the seed from a given plant will always yield that same food. Dark, ominous looking clouds usually precede a fierce storm, and it's prudent to take cover when such clouds are seen. Also, leaves turning color and falling off the trees means that winter is coming, and preparations must be made so as to survive until the following spring.

If we notice that every time it doesn't rain for more than a week our vegetable plants die, we would generate a rule that if there is no rain for a week, we need to irrigate or otherwise somehow water the vegetable garden. Implicit in this is that somewhere a "hypothesis" or "model" is created. In this case our model is that plants need regular watering. When the data are fit to this model, we quantify the case that vegetable plants need water at least once a week, and then the appropriate watering rule may then be created.

An interesting conjecture is that much, if not all, of what we call *the arts* came about because our brains are so interested in seeing patterns that we take delight and often find beauty in well-designed original patterns. Our eyes look at paintings and sculptures, our ears listen to music, our brains process the language constructs of poetry and prose, and so on. In every case we are finding pleasure in studying patterns. Sometimes the patterns are clear, as in a Bach fugue. Sometimes the patterns are harder to recognize, as in a surrealistic Picasso painting. Sometimes we are playing a game looking for patterns that just might not be there—as in a Pollock painting. Perhaps this way of looking at things is sheer nonsense, but then how can you explain how a good book or a good symphony (or rap song if that's your style) or a good painting can grab your attention and in some sense please you? The arts don't seem to be necessary for the basic survival of our species, so why do we have them at all?

A subtle rustling in the brush near the water hole at dusk sometimes—but not always—means that a man-eating tiger is stalking you. It would be to your advantage to make a decision and take action. Even if you're not certain that there's really a tiger present, you should err on the cautious side and beat a hasty retreat; you won't get a second chance. This survival skill is a good example of our evolutionary tendency to look for patterns and to react as if these patterns are there, even when we are not really sure that they indeed are there. In formal terms, you don't have all the data, but you do have *anecdotal* information.

Our prehistoric ancestors lived a very provincial existence. Life spans were short; most people did not live more than about 30 years. They didn't get to see more than about 10,000 sunrises. People outside their own tribe (and possibly some nearby tribes) were hardly ever encountered, so that the average person never saw more than a few hundred other people over the course of a lifetime. Also, very few people (other than members of nomadic tribes) ever traveled more than about 50 miles from where they were born. There are clearly many more items that could be added to this list, but the point has probably been adequately made: Peoples' brains never needed to cope with situations where there were hundreds of thousands or millions of data points to reconcile.

In today's world, however, things are very different: A state lottery could sell a hundred million tickets every few months. There are about six billion (that's six thousand million) people on the earth. Many of us (at least in North America and Western Europe) have traveled thousands of miles from the place of our birth many times; even more of us have seen movies and TV shows depicting places and peoples all over the world. Due to the ease with which people move around, a disease epidemic is no longer a local issue. Also, because we are aware of the lives of so many people in so many places, we know about diseases that attack only one person in a hundred thousand and tragedies that occur just about anywhere. If there's a vicious murderer killing teenage girls in Boston, then parents in California, Saskatoon, and London hear about it on the evening news and worry about the safety of their daughters.

When dealing with unlikely events spread over large numbers of opportunities, your intuition can and does often lead you astray. Since you cannot easily comprehend millions of occurrences, or lack of occurrences, of some event, you tend to see patterns in a small numbers of examples—again the *anecdotal* approach. Even when patterns don't exist, you tend to invent them; you are using your "better safe than sorry" prehistoric evolved response. This could lead to the inability to correctly make many important decisions in your life: What medicines or treatments stand the best chance of curing your ailments? Which proffered medicines have been correctly shown to be useful, and which ones are simply quackery? Which environmental concerns are potentially real and which are simple coincidence? Which environmental concerns are no doubt real but probably so insignificant that it we can reasonably ignore them? Are "sure bets" on investments or gambling choices really worth anything? We need an organized methodology for examining a situation and coping with information, correctly extracting the pattern and the likelihood of an event happening or not happening to us, and also correctly "processing" a large set of data and concluding, when appropriate, that there really is or is not a pattern present.

In other words, we want to understand how to cope with a barrage of information. We need a way of measuring how sure we are of what we know, and when or if what we know is adequate to make some predictions about what's to come.

RANDOM EVENTS AND PROBABILITY

This is a good place to introduce the concepts of random events, random variables, and probability. These concepts will be wrung out in detail in later chapters, so for now let's just consider some casual definitions.

For our purposes an *event* is a particular occurrence of some sort out of a larger set of possible occurrences. Some examples are:

- Will it rain tomorrow? The full set of possible occurrences is the two events Yes—it will rain, and No—it won't rain.
- When you flip a coin, there are two possible events. The coin will either land head side up or tail side up (typically referred to as "heads" or "tails").
- When you roll one die, then there are six possible events, namely the six faces of the die that can land face up—that is, the numbers 1, 2, 3, 4, 5, and 6.
- When you play a quiz game where you must blindly choose "door A, door B, or door C" and there is a prize hiding behind only one of these doors, then there are three possible events: The prize is behind door A, it's behind door B, or it's behind door C.

Variable is a name for a number that can be assigned to an event. If the events themselves are numbers (e.g., the six faces of the die mentioned above), then the most reasonable thing to do is to simply assign the variable numbers to the event numbers. A variable representing the days of the year can take on values 1, 2, 3, ..., all the way up to 365. Both of these examples are of variables that must be integers; that is, 4.56 is not an allowed value for either of them. There are, of course, cases where a variable can take on any value, including fractional values, over some range; for example, the possible amount of rain that fell in Chicago last week can be anything from 0 to 15 inches (I don't know if this is true or not, I just made it up for the example). Note that in this case 4.56, 11.237, or 0.444 are legitimate values for the variable to assume. An important distinction between the variable in this last example and the variables in the first two examples is that the former two variables only can take on a finite number of possibilities (6 in the first case, 365 in the second), whereas by allowing fractional values (equivalently, real number values), there are an infinite number of possibilities for the variable in the last example.

A random variable is a variable that can take on one of an allowed set of values (finite or infinite in number). The actual value selected is determined by a happening or happenings that are not only outside our control but also are outside of any recognized, quantifiable, control—but often do seem to follow some sort of pattern.

A random variable cannot take on any number, but instead must be chosen out of the set of possible occurrences of the situation at hand. For example, tossing a die and looking at the number that lands facing up will give us one of the variables {1, 2, 3, 4, 5, 6}, but never 7, 0, or 3.2.

The most common example of a simple random variable is the outcome of the flip of our coin. Let's assign the number −1 to a tail and +1 to a head. The flip of the coin must yield one of the two chosen values for the random variable, but we seem to have no way of predicting which value it will yield for a specific flip.

Is the result of the flip of a coin truly unpredictable? Theoretically, no: If you carefully analyzed the weight and shape of the coin and then tracked the exact motion of the flipper's wrist and fingers, along with the air currents present and the nature of the surface that the coin lands on, you would see that the flipping of a coin is a totally predictable event. However, since it is so difficult to track all these subtle factors carefully enough in normal circumstances and these factors are extremely difficult to duplicate from flip to flip, the outcome of a coin flip can reasonably be considered to be a random event. Furthermore, you can easily list all the possible values of the random variable assigned to the outcome of the coin flip (−1 or 1); and if you believe that the coin flip is fair, you conclude that either result is equally likely. This latter situation isn't always the case.

If you roll two dice and define the random variable as the sum of the numbers you get from each die, then this random variable can take on any value from 2 to 12. All of the possible results, however, are no longer equally likely. This assertion can be understood by looking at every possible result as shown in Table 1.1.

As may be seen from the table, there is only one way that the random variable can take on the value 2: Both dice have to land with a 1 face up. However, there are three ways that the random variable can take on the value 4: One way is for the first die to land with a 1 face up while the second die lands with a three face up. To avoid writing this out over and over again, I'll call this case {1, 3}. By searching through the table, we see that the random variable value of 4 can be obtained by the dice combinations {1, 3}, {2, 2}, and {3, 1}.

I'll create a second table (Table 1.2) that tabulates the values of the random variable and the number of ways that each value can result from the rolling of a pair of dice:

The numbers in the right-hand column add up to 36. This is just a restatement of the fact that there are 36 possible outcomes possible when rolling a pair of dice.

Define the *probability* of a random event as the number of ways that that event can occur, divided by the number of all possible events. Adding a third column to the table to show the probabilities, I get Table 1.3.

For example, if you want to know the probability that the sum of the numbers on the two dice will be 5, the second column of this table tells us that there are four ways to get 5. Looking back at the first table, you can see that this comes about from the possible combinations {1, 4}, {2, 3}, {3, 2} and {4, 1}. The probability of rolling two dice and getting a (total) of 5 is therefore 4/36,

TABLE 1.1. All the Possible Results of Rolling a Pair of Dice

First Die Result	Second Die Result	Random Variable Value = Sum of First & Second Results	First Die Result	Second Die Result	Random Variable Value = Sum of First & Second Results
1	1	2	4	1	5
1	2	3	4	2	6
1	3	4	4	3	7
1	4	5	4	4	8
1	5	6	4	5	9
1	6	7	4	6	10
2	1	3	5	1	6
2	2	4	5	2	7
2	3	5	5	3	8
2	4	6	5	4	9
2	5	7	5	5	10
2	6	8	5	6	11
3	1	4	6	1	7
3	2	5	6	2	8
3	3	6	6	3	9
3	4	7	6	4	10
3	5	8	6	5	11
3	6	9	6	6	12

TABLE 1.2. Results of Rolling a Pair of Dice Grouped by Results

Value of Random Variable	Number of Ways of Obtaining this Value
2	1
3	2
4	3
5	4
6	5
7	6
8	5
9	4
10	3
11	2
12	1

TABLE 1.3. Same as Table 1.2 but also Showing Probability of Results

Value of Random Variable	Number of Ways of Obtaining this Result	Probability of Getting this Result
2	1	1/36 = 0.028
3	2	2/36 = 0.056
4	3	3/36 = 0.083
5	4	4/36 = 0.111
6	5	5/36 = 0.139
7	6	6/36 = 0.167
8	5	5/36 = 0.139
9	4	4/36 = 0.111
10	3	3/36 = 0.083
11	2	2/36 = 0.056
12	1	1/36 = 0.028

sometimes called "4 chances out of 36." 4/36 is of course the same as 2/18 and 1/9 and the decimal equivalent, 0.111.[1]

If you add up all of the numbers in the new rightmost column, you'll get exactly 1. This will always be the case, because it is the sum of the probabilities of all possible events. This is the "certain event" and it must happen; that is, it has a probability of 1 (or 100%). This certain event will be that, when you toss a pair of dice, the resulting number—the sum of the number of dots on the two faces that land face up—again must be some number between 2 and 12.

Sometimes it will be easier to calculate the probability of something we're interested in *not* happening than to calculate the probability of it happening. In this case since we know that the probability of our event either happening or not happening must be 1, then the probability of the event happening is simply 1—the probability of the event not happening.

From Table 1.3 you can also calculate combinations of these probabilities. For example, the probability of getting a sum of *at least 10* is just the probability of getting 10 + the probability of getting 11 + the probability of getting 12, = 0.083 + 0.056 + 0.028 = 0.167. Going forward, just for convenience, we'll use the shorthand notation Prob(12) to mean "the probability of getting 12," and we'll leave some things to the context; that is, when rolling a pair of dice, we'll assume that we're always interested in the sum of the two numbers facing up, and we'll just refer to the number.

Exactly what the probability of an event occurring really means is a very difficult and subtle issue. Let's leave this for later on, and just work with the

---

[1] Many fractions, such as 1/9, 1/3, and 1/6, do not have exact decimal representations that can be expressed in a finite number of digits. 1/19, for example, is 0.111111111 …, with the 1's going on forever. Saying that the decimal equivalent of 1/9 is 0.111 is therefore an approximation. Knowing how many digits are necessary to achieve a satisfactory approximation is context-dependent—there is no easy rule.

intuitive "If you roll a pair of dice very many times, about 1/36 of the time the random variable will be 2, about 2/36 of the time it will be 3, and so on."

An alternative way of discussing probabilities that is popular at horse races, among other places, is called *the odds* of something happening. Odds is just another way of stating things. If the probability of an event is 1/36, then we say that the odds of the event happening is 1 to 35 (usually written as the ratio 1:35). If the probability is 6/36, then the odds are 6:30 or 1:5, and so on. As you can see, while the *probability* is the number of ways that a given event can occur divided by the total number of possible events, the *odds* is just the ratio of the number of ways that a given event can occur to the number of ways that it can't occur. It's just another way of expressing the same calculation; neither system tells you any more or less than the other.

In the simple coin flip game, the probability of winning equals the probability of losing,=0.5. The odds in this case is simply 1:1, often called *even odds*. Another *term of art* is the case when your probability of winning is something like 1:1000. It's very unlikely that you'll win; these are called *long odds*.

Something you've probably noticed by now is that I tend to jump back and forth between fractions (such as 1/4) and their decimal equivalents (1/4 = 0.25). Mathematically, it doesn't matter which I use. I tend to make my choice based on context: When I want to emphasize the origins of the numerator and denominator (such as 1 chance out of 4), I'll usually use the fraction, but when I just need to show a number that's either the result of a calculation or that's needed for further calculations, I'll usually use the decimal. I hope this style pleases you rather than irritates you; the important point is that insofar as the mathematics is concerned, both the fraction and the decimal are equivalent.

You now have the definitions required to look at a few examples. I'll start with some very simple examples and work up to some fairly involved examples. Hopefully, each of these examples will illustrate an aspect of the issues involved in organizing some probabilistic data and drawing the correct conclusion. Examples of statistical inference will be left for later chapters.

THE LOTTERY {VERY IMPROBABLE EVENTS AND VERY LARGE DATA SETS}

Suppose you were told that there is a probability of 1 in 200 million (that's 0.000000005 as a decimal) of you getting hit by a car and being seriously injured or even killed if you leave your house today. Should you worry about this and cancel your plans for the day? Unless you really don't have a very firm grip on reality, the answer is clearly *no*. There are probabilities that the next meal you eat will poison you, that the next time you take a walk it will start storming and you'll be hit by lightening, that you'll trip on your way to the bathroom and split your skull on something while falling, that an airplane will fall out of the sky and crash through your roof, and so on. Just knowing that you and your acquaintances typically do make it through the day is anecdotal evidence that the sum of these probabilities can't be a very large number. Looking at your city's accidental death rate as a fraction of the total population gives you a pretty realistic estimate of the sum of these probabilities. If you let your plans for your life be compromised by every extremely small probability of something going wrong, then you will be totally paralyzed.[2] One in two hundred million, when it's the probability of something bad happening to you, might as well be zero.

Now what about the same probability of something good happening to you? Let's say you have a lottery ticket, along with 199,999,999 other people, and one of you is going to win the grand prize. Should you quit your job and order a new car based on your chance of winning?

The way to arrive at an answer to this question is to calculate a number called the expected value (of your winnings). I'll define expected value carefully in the next chapter, but for now let me just use the intuitive "What should I expect to win?" There are 4 numbers I need in order to perform the calculation.

First, I need the probability of winning. In this case it's 1 in 200 million, or 0.000000005. Next, I need the probability of losing. Since the probability of losing plus the probability of winning must equal 1, the probability of losing must be $1 - 0.000000005 = .999999995$.

I also need the amount of money you will make if you win. If you buy a lottery ticket for $1 and you will get $50,000,000 if you win, this is $50,000,000 - \$1 = \$49,999,999$.

Lastly, I need the amount of money you will lose if you don't win. This is the dollar you spent to buy the lottery ticket. Let's adopt the sign convention that winnings are a positive number but losses are a negative number. The amount you'll lose is therefore −$1.

In order to calculate the expected value of your winnings, I add up the product of each of the possible money transfers (winning and losing) multiplied by the probability of this event. Gathering together the numbers from above, we obtain

$$\text{Expected value} = (0.000000005)(\$49,999,999) - (.999999995)(\$1)$$
$$\approx (0.000000005)(\$50,000,000) - (1)(\$1) = \$0.25 - \$1.00 = -\$0.75$$

I have just introduced the symbol "$\approx$", which means "not exactly, but a good enough approximation that the difference is irrelevant." "Irrelevant," of course, depends on the context of the situation. In this example, I'm saying

---

[2] In 1976, when the U.S. Skylab satellite fell from the sky, there were companies selling Skylab insurance—coverage in case you or your home got hit. If you consider the probability of this happening as approximately the size of the satellite divided by the surface area of the earth, you'll see why many fortunes have been made based on the truism that "there's a sucker born every minute."

that $(0.000000005)(\$49,999,999) = \$0.249999995$ is close enough to $0.25 that when we compare it to $1.00 we never notice the approximation.

The expected value of your winnings is a negative number—that is, you should expect to lose money. What the expected value is actually telling you is that if you had bought *all* of the lottery tickets, so that you had to be the winner, you would still lose 75 cents on every dollar you spent. It's no wonder that people who routinely calculate the value of investments and gambling games often refer to lotteries as a "Tax on Stupidity."

What I seem to be saying so far is that events with extremely low probabilities simply don't happen. If we're waiting for you to win the lottery, then this is a pretty reasonable conclusion. However, the day after the lottery drawing there will be an article in the newspaper about the lottery, along with a picture of a very happy person holding up a winning lottery ticket. This person just won 50 million dollars!

Am I drawing two different conclusions from the same set of data? Am I saying both that nobody wins the lottery and that somebody always wins the lottery? The answer is that there is no contradiction, we just have to be very careful how we say what we say. Let me construct an example. Suppose the state has a lottery with the probability of any one ticket winning = 0.000000005 and the state sells 200 million tickets, which include every possible choice of numbers. It's an absolute certainty that *somebody* will win (we'll ignore the possibility that the winning ticket got accidentally tossed into the garbage). This does not at all contradict the statement that it's "pretty darned near" certain that *you* won't win.

What we are struggling with here is the headache of dealing with a very improbable event juxtaposed on a situation where there are a huge number of opportunities for the event to happen. It's perfectly reasonable to be assured that something will never happen to you while you know that it will happen to somebody. Rare diseases are an example of this phenomenon. You shouldn't spend much time worrying about a disease that randomly afflicts one person in, say, 10 million, every year. But in the United States alone there will be about 30 cases of this disease reported every year, and from a Public Health point of view, somebody should be paying attention to it.

A similar situation arises when looking at the probability of an electrical appliance left plugged in on your countertop starting a fire. Let's say that this probability is 1 in 30,000 per person.[3] Should you meticulously unplug all your countertop kitchen appliances when you're not using them? Based on the above probability, the answer is "don't bother." However, what if you're the senior fire department safety officer for New York City, a city with about 8 million residents? I'll assume an average of about 4 people per residence. If

---

[3] The U.S. Fire Administration's number is about 23,000 appliance related electrical fires per person. I rounded this up to 30,000 to make a convenient comparison to a population of about 300 million.

nobody unplugs their appliances, then you're looking at about 20 unnecessary fires every year, possible loss of life, and certain destruction of property. You are certainly going to tell people to unplug their appliances. This is a situation where the mathematics might not lead to the right answer, assuming that there is a right answer. You'll have to draw your own conclusion here.

One last note about state lotteries. In this example, the state took in $200 million and gave out $50 million. In principle, this is why state lotteries are held. The state makes money that it uses for education or for health care programs for the needy, and so on. From this perspective, buying a lottery ticket is both a social donation and a bit of entertainment for you—that's not a bad deal for a dollar or two. On the other hand, as an investment this not only has an absurdly low chance of paying off, but since the expected value of the payoff is negative, this is not what's called a *Fair Game*. From an investment point of view, this is a very poor place to put your money.

COIN FLIPPING {FAIR GAMES, LOOKING BACKWARDS FOR INSIGHT}

Let's set up a really simple coin flipping game between you and a friend. One of you flips a coin. If the result is heads, you collect a dollar from your friend; if the result is tails, you pay a dollar to your friend. Assuming that the coin is fair (not weighted, etc.), there is an equal probability of getting either a head or a tail. Since the sum of all the probabilities must equal one, then the probability of getting a head and the probability of getting a tail must be equal to 0.5 (one-half).

Letting +$1 be the result of winning (you get to be a dollar richer) and letting −$1 be the result of losing (you get to be a dollar poorer), then the expected value of your return is

$$E = (0.5)(+1\$) + (0.5)(-1\$) = 0$$

This is what I'm calling a fair game. Since I am defining positive $ values as winning (money coming to you) and defining negative $ values as losing (money leaving you), then if your winnings and your losings exactly balance out, the algebraic sum is zero. Nobody involved (in this case there is just you and your friend) should *expect* to win, or lose, money. This last sentence seems at first to be very simple. Examining the nuances of what is meant by *expect* thoroughly, however, will take a significant portion of this book.

While Expected Value is a mathematical term that is carefully defined, the definition is not quite the same as the common, conversational, use of the term *expected* or *expectation*. Also, as a few examples will show, in most cases it is much less likely that you will come away from this game with the exact amount of money that you went into the game with than that you either win or lose some money.

The simplest example of this last claim is a game where you flip the coin once and then stop playing. In this case you must either win or lose a dollar—there are no other choices. For that matter, if the game is set up so that there will be any odd number of coin flips, it is impossible for you to come away neither winning nor losing. In these cases it is impossible to win the *expected* amount of winning—clearly a distinction between the mathematical definition of Expected Value and its common usage.

Now let's look at some games where there are an even number of coin flips. In these cases it certainly is possible to end up with a zero sum. The simplest example is the two-flip game. If you flip first a head and then a tail (or vice versa), you come away with a zero sum.

As a brief aside, let's introduce some mathematical notation that will make things easier as we proceed. If we let the letter n refer to the number of coin flips, then our two-flip game can be referred to as an $n = 2$ game. The choice of the letter n was completely arbitrary. No new ideas or calculations have just been presented. This is simply a convenient way to talk about things. At this point there is very little to be gained by doing this, but the approach will prove to be very powerful and useful as things get more complicated.

Returning to the examination of coin flipping, let's examine the $n = 2$ game in detail. This is easy to do because there are only four possible scenarios. I'll show both the results of the flip (heads or tails) and the algebraic value of the random variable associated with these results, just for clarity in Table 1.4.

As the table shows, there are two opportunities for a zero sum: one opportunity for a positive sum and one opportunity for a negative sum. The probability that you will win is therefore 0.25, the probability that you'll lose is 0.25, and the probability that you'll break even is 0.5. In other words, it's equally likely that there will be a zero sum and that there will be a nonzero sum.

Now let's look at a game with a larger value of n. What about, say, $n = 10$? How many table entries must there be? For each flip of the coin, there are two possibilities. Therefore, for $n = 2$ there are 4 possibilities, for $n = 3$ there are $(2)(2)(2) = 8$ possibilities, for $n = 4$ there are $(2)(2)(2)(2) = 16$ possibilities, and so on. Adding to our mathematical notation toolbox, the expression 2^n means 2 multiplied by itself n times. In other words, $2^3 = 8$, $2^4 = 16$, and so on. We can therefore say that "There are 2^n possibilities for an n-flip coin flip game."

A 10-flip game would have 2^{10} possible outcomes. Working this out, we obtain

TABLE 1.4. All the Possible Results of an $n = 2$ Coin Flip Game

First Flip	First Flip Variable	Second Flip	Second Flip Variable	Sum
Head	+1	Tail	−1	0
Head	+1	Head	+1	2
Tail	−1	Tail	−1	−2
Tail	−1	Head	+1	0

$$2^{10} = 1024$$

While there is no reason why I couldn't create the list and examine every possible outcome, it certainly does not look like doing this would be fun. Let's set our goals a little more humbly, I'll look at $n=3$ and $n=4$ games. Adding to our notational soup mix, let the letter k refer to a particular flip; that is, $k=2$ refers to the second flip, $k=3$ refers to the third flip, and so on. Clearly, in an nth-order game, k can take on all of the (integer) values from 1 to n. In algebraic notation, this is written as

$$1 \leq k \leq n$$

The symbol $\leq$ means "less than or equal to," so the above expression is read as "1 is equal to or less than k, and also k is equal to or less than n." It may not seem that way now, but this is actually a very convenient way to express things.

Using our new notational ability, Table 1.5 shows the $n=3$ game.

As expected, there is absolutely no way to play an $n=3$ game and come out even; you have to either win or lose some amount. However, since each of the outcomes (rows) in the above table is equally likely, and there are 8 of them, the probability of each outcome is exactly 1/8, so that the expected value of the your return is

$$\frac{-3}{8} + \frac{-2}{8} + \frac{-1}{8} + \frac{+1}{8} + \frac{-1}{8} + \frac{+1}{8} + \frac{+2}{8} + \frac{+3}{8} = \frac{0}{8} = 0$$

The expected value is zero, so this is indeed a fair game. Worth noting again in this case is the fact that the expected value might be a value that you can never realize. You can play $n=3$ games all night, and you will never come away with a 0 (no loss or no gain) result from one of these games. What you

TABLE 1.5. All the Possible Results of an $n=3$ Coin Flip Game

$K=1$	$k=2$	$k=3$	Sum
−1	−1	−1	−3
−1	−1	+1	−2
−1	+1	−1	−1
−1	+1	+1	+1
+1	−1	−1	−1
+1	−1	+1	+1
+1	+1	−1	+2
+1	+1	+1	+3

might expect, however, is that after a night of $n=3$ games, the average of your results would be close to zero.

In this simple coin flip situation, a hundred $n=3$ games is exactly the same as an $n=300$ game. We can therefore extend our logic: For any number of $n=$ anything games, if the total number of coin flips is odd, you can still never walk away with a zero sum. You might, however, come close.

Let's look at an $n=4$ game and see what happens (Table 1.6):

In an $n=4$ game, there are $2^4 = 16$ possible outcomes, all listed in Table 1.6. Looking through the column of sums, we see that there are 6 possible ways to get 0. In other words, the probability of neither winning nor losing is $6/16 = .375$. This is lower than the probability of neither winning nor losing was for an $n=2$ game. Is this a pattern?

Table 1.7 shows that the probability of getting a zero sum out of even order games for $2 \leq n \leq 20$.

There is indeed a trend: As n gets larger, the probability of getting a zero sum gets lower. In other words, the more times you flip the coin, the less likely it is that you will get exactly the same number of heads and tails.

TABLE 1.6. All the Possible Results of an $n=4$ Coin Flip Game

$k=1$	$k=2$	$k=3$	$k=4$	Sum	$k=1$	$k=2$	$k=3$	$k=4$	Sum
−1	−1	−1	−1	−4	1	−1	−1	−1	−2
−1	−1	−1	1	−2	1	−1	−1	1	0
−1	−1	1	−1	−2	1	−1	1	−1	0
−1	−1	1	1	0	1	−1	1	1	2
−1	1	−1	−1	−2	1	1	−1	−1	0
−1	1	−1	1	0	1	1	−1	1	2
−1	1	1	−1	0	1	1	1	−1	2
−1	1	1	1	2	1	1	1	1	4

TABLE 1.7. Probabilities of 0 Sum for $n=2$ to $n=20$ Coin Flip Games

n	Probability of 0 Sum
2	0.500
4	0.375
6	0.313
8	0.273
10	0.246
12	0.226
14	0.209
16	0.196
18	0.185
20	0.176

This leads to some very thought-provoking discussions of just what a probability means, what you can expect, and how sure you can be of what you expect. This, in turn, leads to a very basic question of just what it means to be sure about something, or in other words, just how confident you are about something happening. In the Chapter 3 I'll define a "confidence factor" by which we can gauge just how sure we are about something.

Returning to the simple coin flip, what if I want the probability of getting 5 heads in a row? This is identically the probability of flipping 5 coins and getting 5 heads, because the results of coin flips, be they with multiple coins or with the same coin over and over again, are independent of each other. You could get the answer by writing a list such as the tables I've been presenting for the $n = 5$ case or you could just note that for independent events, all you have to do is to multiply the individual probabilities together. In other words, the probability of getting 5 heads in a row is just $(1/2)^5 = 1/32 \approx 0.033$.

1/32 is a low probability, but not so low as to astound you if it happens. What if it indeed just did happen? You flipped a coin and got 5 heads in a row. What's the probability of getting a head if you now flip the coin again? Assuming, of course, that the coin isn't weighted or corrupted in any other manner (i.e., that the flips are indeed fair), the probability of a head on this flip (and for any subsequent flip) is still just 1/2. Putting it simply, a coin has no memory.

Reiterating the point above, flipping one coin 6 times is statistically identical to flipping six different coins once each and then examining the results. It doesn't matter whether you flip the six coins one at a time or if you toss them all up into the air and let them fall onto the table. The six coins are independent of each other: They do not "know" or "remember" anything about either their own past performance or the performance of any other coin. When you look at it this way, it's pretty clear that the flip of the sixth coin has nothing to do with the flips of the first five coins. For that matter, if you tossed all six coins into the air at once, you couldn't even say which coin is the "sixth coin."

The above arguments are a simple case of another interesting discussion: When does knowledge of past results tell you something about future results? In the above example, it doesn't tell you anything at all. Later in this chapter I will show an example where this isn't the case.[4]

Returning to the coin flipping game, remember that the expected value of return of an $n = $ anything coin flip is always zero, as has been illustrated in several examples. If you were to flip a coin 10 times, the most likely single result would be an equal number of heads and tails, even though it's not a very likely event (remember, in this case we're counting the number of ways

---

[4] People who don't understand this point will tell you that if you flipped a coin 100 times and got 100 heads, you should bet on a tail for the 101st flip because "it's due." I'd be more inclined to bet on a 101st head, because after the first 100 heads I'd be pretty sure that I wasn't dealing with a fair coin.

of getting 5 heads and 5 tails out of $2^{10} = 1024$ possible configurations). The distinction between *most* likely and yet *not very* likely (or the equivalent, very unlikely) eludes many people, so let's consider another example.

Suppose I have a giant roulette wheel with 1000 slots for the ball to land in. I'll number the slots 1 to 999 consecutively, and then number the thousandth slot 500. This means I have exactly one slot for each number between 1 and 999, except for the number 500, for which I have two slots. When I spin this roulette wheel and watch for the ball to settle in a slot, I see that there are two opportunities for the ball to settle in a slot numbered 500, but only one opportunity for the ball to settle at any other number. In other words, the probability of the ball landing at 500 is twice the probability of the ball landing at any other number. 500 is clearly the most likely result. The probability of the ball landing in a 500 slot is 2 out of 1000, or 0.002. The probability of the ball *not* landing in a 500 slot—that is, the probability of landing *anywhere but* in a 500 slot—is 998 out of 1000, or 0.998. It is, therefore, very unlikely that the ball will land in a 500 slot. Now let's combine both of the above observations into the same sentence: The most likely slot the ball will land in will be numbered 500, but it is very unlikely that the ball will land in a slot numbered 500 as compared to some other number.

Returning to the coin flipping example, no matter how many (even number of) times you flip a coin, the most likely result is that you'll get an equal number of heads and tails. The more times you flip the coin, however, the less likely this result will be. This same idea will be presented in the chapter on random walks.

A variation on the above example is the picking of a number for a lottery. If you were to pick, say, 12345, or 22222, you would be criticized by the "experts": "You never see a winning number with a regular pattern—it's always something like 13557 or 25738 or.... " This last statement is correct. It is correct because of all the nearly one million five-digit numbers that can be picked, very few of them have simple, recognizable digit patterns. It is therefore most likely that the winning number will *not* have a recognizable pattern. However, the five lottery balls have no memory or awareness of each other. They would not "know" if they presented a recognizable pattern. Any five-digit number is equally likely. The difference between the recognizable patterns and other patterns is only in the eyes of the beholder, I can't even imagine how I'd define a "recognizable" pattern except maybe by the majority vote of a room full of people.

A corollary to this is of course that there's no reason not to pick the very number that won last week. It's highly unlikely that this number will win again just because there are so many numbers to choose from, but it is just as likely that this number will win as it is that any other number will win.

Moving on, what if you have just flipped a coin five times, got five heads, and now want to flip the coin ten times more? The expectation value looking forward is still zero. But, having just won the game five times, you have five dollars more in your pocket than you started with. Therefore, the most likely scenario is that you will end up with five dollars in your pocket! This property will be covered in more detail in the chapter on gambling games (Chapter 8). Generalizing this conclusion, I can say that if you are going to spend the evening flipping coins, your most likely status at the finish is just your status at the time you think about it. If you start off lucky (i.e., have net winnings early on), then you'll probably end up winning a bit, and vice versa. There really is such a thing as "being on a winning streak." However, this observation can only be correctly made after the fact. If you were lucky and got more heads than tails (or vice versa, if that's the side you're on), then you were indeed on a winning streak. The perception that being on a winning streak so far will influence the coin's future results is of course total nonsense. You might win a few times in a row, you might even win most of the time over the course of the evening, but each flip is still independent of all the others.

There is an argument for saying that if you have been on a winning streak, it's more likely that you'll end the evening ahead (i.e., with net winnings rather than losses) than if you haven't been on a winning streak. That argument is that if you have been on a winning streak, you have a lot more money in your pocket than you would have if you had been on a losing streak. You are therefore in a better position to withstand a few (more) losses without being wiped out and having to quit playing, and therefore your odds of winning for the evening have been increased. This argument has nothing to do with the probabilities of an individual win (coin flip, roulette wheel, poker hand, whatever). If you are just playing for a score on a piece of paper and cannot be "wiped out," this argument is worthless.

THE COIN FLIP STRATEGY THAT CAN'T LOSE

Assume that you want to earn $1 a day (scale this to $10, or $100, or anything you wish—the discussion is clearest when working with a starting value of $1). Let's build a strategy for playing a clever coin flipping game:

1. Bet $1 on the results of a coin flip.
2. If you win the first coin flip, you've earned your $1 for the day. Go home.
3. If you lose the first coin flip, you've lost $1. Bet $2 and try again.
4. If you win the second coin flip, then you've recovered your lost $1 and won $1. Go home.
5. If you lose the second coin flip, then you've now lost $3. Bet $4 and try again.
6. If you win the third coin flip, then you've recovered your lost $7 and won $1. Go home.
7. If you lose the third coin flip, then you've now lost $7. Bet $8 and try again.
8. And so on, until you win.

This scheme seems unbeatable. If you keep flipping a coin, sooner or later you have to get a head, and you've won for the day. What could possibly go wrong?

The scheme could be analyzed in detail from a number of different perspectives. The Achilles' heel is that you need a very big wallet. For example, to cover three losses and still have the money to place your fourth bet, you need to start with $1 + $2 + $4 + $8 = $15. In general, to be able to place n bets you need to start out with $2^n - 1$ dollars.

If you start the evening with 1 million dollars, it's pretty certain that you will be able to go home with $1 million + $1. You have enough money to cover a lot of tails until you get your head. On the other hand, if you show up with only $1, then the probability of you going home with your starting $1 plus your winnings of $1 is only 50%. Putting this last case slightly differently, the probability of you doubling your money before getting wiped out is 50%. Without showing the details now, let me just say that it turns out that no matter now much money you start with, the probability of you doubling your money before you get wiped out is at best 50%, less if you have to flip the coin many times. If you're "earning" only a dollar a day, then you need to come back the number of days equal to the number of dollars you're starting with. You can save yourself a lot of time, however, by just betting all of your money on the first coin flip—a very simple game with a 50% probability of doubling your money and a 50% probability of being wiped out. Here again, we have been reconciling an unlikely event (not getting a head after many flips of a coin) with a large number of opportunities for the event to happen (many flips of a coin). We should note here that it's not possible to "fool Mother Nature." If you start out with a million dollars and plan to visit 10,000 gambling houses each night, hoping to win only $1 at each house, then the probabilities start catching up with you and you no longer have a sure thing going.

THE PRIZE BEHIND THE DOOR {LOOKING BACKWARDS FOR INSIGHT, AGAIN}

This example is subtle, and the answer is often incorrectly guessed by people who should know better. There are still ongoing debates about this puzzle on various Probability-Puzzle websites because the correct answer seems to be so counterintuitive to some people that they just won't accept the analysis. It is known by many names, perhaps most commonly the "Monty Hall" problem, named after the host of a TV game show.

You are a participant in a TV game show. There are three doors (let's call them doors A, B, and C). Behind one of these doors is a substantial prize, behind the other two doors is nothing. You have to take a guess. So far this is very straightforward: Your probability of guessing correctly and winning the prize is exactly 1/3.

You take (and announce) your guess. Before the three doors are opened to reveal the location of the prize, the game show host goes to one of the two doors that you *didn't* choose, opens it, and shows you that the prize is *not* behind this door. The prize, therefore, must either be behind the unopened door that you chose or behind the unopened door that you did not choose. You are now given the option of staying with your original choice or switching to the unopened door that you did not choose. What should you do?

Almost everybody's first response to this puzzle is to shrug—after all, there are now two unopened doors and the prize is behind one of them. Shouldn't there simply be a 0.5 (50%) probability of the prize being behind either of these doors, and therefore it doesn't matter whether you stay with your original choice or switch?

Let's look at all the possible scenarios. Assume that that your first guess is door B. (It doesn't matter which door you guess first, the answer always comes out the same.)

1. If the prize is behind door A, then the host must tell you that the prize is not behind door C.
2. If the prize is behind door B, then the host can tell you either that the prize is not behind door A or that the prize is not behind door C.
3. If the prize is behind door C, then the host must tell you that the prize is not behind door A.

Since each of the above three situations is equally likely, they each have a probability of 1/3.

In situation 1, if you stay with your first choice (door B), you lose. You have the option of switching to door A. If you switch to door A, you win.

In situation 2, if you stay with your first choice (door B), you win. If the host tells you that the prize is not behind door A and you switch to door C, you lose. Also, if the host tells you that the prize is not behind door C and you switch to door A, you lose.

In situation 3, if you stay with your first choice (door B), you lose. You have the option of switching to door C. If you switch to door C, you win.

At this point, Table 1.8 is in order. Remember, your first choice was door B.

TABLE 1.8. Monty Hall Game, Door B Being the First Choice

Prize Location	Remaining Doors	Definite Losers	Stay With Choice B	Switch
A	A, C	C	Lose	Win
B	A, C	A & C	Win	Lose
C	A, C	A	Lose	Win

It appears that if you stay with your first choice, you only win in one of three equally likely situations, and therefore your probability of winning is exactly 1/3. This shouldn't really surprise you. The probability of correctly guessing one door out of three is 1/3, and there's not much more that you can say about it.

On the other hand, if your only options are to stay with your first choice or to switch to the other unopened door, then your probability of winning if you switch must be $1 - 1/3 = 2/3$. There's no getting around this: Either you win or you lose and the probability of winning plus the probability of losing must add up to the certain event—that is, to a probability of 1.

What just happened? What has happened that's different from having just flipped a coin five times, having gotten five heads, and wondering about the sixth flip?

In the coin flipping example, neither the probabilities of the different possibilities or your knowledge of these probabilities changed after five coin flips. In other words, you neither changed the situation nor learned more about the situation. (Obviously, if someone took away the original coin and replaced it with a two headed coin, then expectation values for future flips would change.) In this game-show example, only your knowledge of the probabilities changed; you learned that the probability of the prize being behind one specific door was zero. This is enough, however, to make it possible that the expected results of different actions on your part will also change.

In a later chapter we'll look at another very unintuitive situation: a combination of games, known as "Parrondo's Paradox," where jumping randomly between two losing games creates a winning game because one of the losing games involves looking back at how much you've already won or lost.

THE CHECKERBOARD {DEALING WITH ONLY PART OF THE DATA SET}

Imagine an enormous checkerboard: The board is 2000 squares wide by 2000 squares long. There are $2000 \times 2000 = 4,000,000$ (four million) squares on the board. Assume that each square has an indentation that can capture a marble.

I'll treat the board as an imaginary map and divide it up into regions, each region containing 1000 indentations. The regions themselves need not be square or even rectangular, so long as each region contains exactly 1000 indentations. There are $4,000,000/1000 = 4000$ of these regions on the board. There is nothing magic in the choice of any of these numbers. For the purposes of the example, I simply need a large total area (in this case 4 million squares) divided into a lot of small regions (in this case 4000 regions). Also, the regions do not all have to be the same size, it just makes the example easier to present.

Now, I'll lay this checkerboard flat on the ground and then climb up to the roof of a nearby building. The building must be tall enough so that the checkerboard looks like a small dot when I look down. This is important because it assures that if I were to toss a marble off the roof, it would land randomly somewhere on or near the checkerboard, but that I can't control where. I then start tossing marbles off the roof until 40,000 marbles have landed on the checkerboard and are trapped in 40,000 indentations. This is, admittedly, a very impractical experiment. I won't worry about that, however, because I'm not really planning to perform the experiment. I just want to describe a way of picturing the random scattering of 40,000 objects into 4,000,000 possible locations. The choice of 40,000 objects isn't even a critical choice, it was just important to choose a number that is a small fraction of 4,000,000 but is still a fairly large number of objects. In any case, when I am done we see that the fraction of the number of indentations that I have filled is exactly

$$\frac{40,000}{4,000,000} = \frac{1}{100} = 0.01 = 1\%$$

Now let's take a close look at a few of the 4000 regions, each of which has 1000 indentations. Since 1% of the indentations are filled with marbles, we would expect to see 1% of 1000, or

$$0.01 \times 1000 = 10$$

marbles in each region. On the average, over the 4000 regions, this is exactly what we must see—otherwise the total number of marbles would not be 40,000. However, when we start looking closely, we see something very interesting[5]: Only about 500 of the regions have 10 marbles.[6] About 200 of the regions have 14 marbles, and about 7 of the regions have 20 marbles. Also, about 9 of the regions have only 2 marbles. Table 1.9 tabulates these results.

What Table 1.9 is showing us is that while the most likely situation, in this case 10 marbles per region, will happen more often than any other situation, the most likely situation is not the only thing that will happen (just like in the coin flip game). The results are distributed over many different situations, with less likely situations happening less often. In order to predict what we see from this experiment, therefore, we not only need to know the most likely result, but also need to know something about how a group of results will be distributed among all possible results. These *probability distributions* will be a subject of the next chapter.

---

[5] At this point I am not attempting to explain how the observations of "what we actually see" come about. This will be the subject of the chapter on binomial distributions (Chapter 6).
[6] I say "about" because it is very unlikely that these numbers will repeat exactly if I were to clear the board and repeat the experiment. How the results should be expected to vary over repeated experiments will also be the subject of a later chapter.

TABLE 1.9. Expected Number of Marbles in Regions on Giant Checkerboard

Number of Regions	Number of Marbles	Number of Regions	Number of Marbles
2	1	292	13
9	2	208	14
30	3	138	15
74	4	86	16
150	5	50	17
251	6	28	18
360	7	14	19
451	8	7	20
502	9	3	21
503	10	2	22
457	11	1	23
381	12		

Before leaving this example, let's play with the numbers a little bit and see what we might learn. Since the most likely result (10 marbles per region) occurs only about 500 times out of 4000 opportunities, some other results must be occurring about 3500 times out of these 4000 opportunities. Again, we have to be very careful what we mean by the term "most likely result." We mean the result that will probably occur *more times than any other result* when we look at the whole checkerboard. The probability that the most likely result will not occur in any given region is about

$$\frac{3500}{4000} = \frac{7}{8} = 0.875 = 87.5\%$$

Putting this in gambling terms, there are 7 to 1 odds against the most likely result occurring in a given region.

Now, suppose someone is interested in the regions that have at least 20 marbles. From the table we see that there are 13 of these regions. It wouldn't be surprising if a few of them are near an edge of the board. Let's imagine that this person locates these regions and takes a picture of each of them. If these pictures were shown to you and you are not able to look over the rest of the board yourself, you might tend to believe the argument that since there are some regions near the edge of the board that have at least twice the average number of marbles, then there must be something about being near the edge of the board that "attracts marbles." There are, of course, many regions that have less than 1/2 the average number of marbles, and some of these are probably near the edge of the board too, but this information is rarely mentioned (it just doesn't make for good headlines). Instead, we see "Cancer Cluster Near Power Lines" and similar statements in the newspapers.

It's hard to generalize as to whether the reporters who wrote the story intentionally ignored some data, unintentionally overlooked some data, didn't understand what they were doing by not looking at all the data, or were just looking to write a good story at the cost of total, complete truthfulness.

In all fairness, I am now leaving the realm of mathematics and meddling in the realms of ecology, public health, and so on. There are indeed unfortunate histories of true disease clusters near waste dumps, and so on. The point that must be made, over and over again, however, is that you cannot correctly spot a pattern by "cherry picking" subsets of a large data set. In the case of power lines near towns, when you look at the entire checkerboard (i.e., the entire country), you find high disease clusters offset by low disease clusters and when you add everything up, you get the average. If there were high disease clusters which were truly nonrandom, then these clusters would not be offset by low disease clusters, and you would get a higher-than-average disease rate when you add everything up. Also, you must be prepared to predict the variability in what you see; for example, a slightly higher-than-average total might just be a random fluctuation, sort of like flipping a coin and getting 5 heads in a row. The World Health Organization maintains a database on their website in the section on Electromagnetic Fields, where you can get a balanced perspective of the power-line and related studies.

This same issue shows up over and over again in our daily lives. We are bombarded with everything from health food store claims to astrological predictions. We are never shown the results of a large study (i.e., the entire checkerboard). The difficult part here is of course that some claim might be absolutely correct. The point, however, is that we are not being shown the information necessary to see the entire picture, so we have no way of correctly concluding whether or not a claim is correct based upon looking at the entire picture or just anecdotal without further investigation. And, of course in the area of better health or longer life or . . . , we are genetically programmed to bet on the pattern we are shown, just in case there are man-eating tigers lurking in the brush.

It is not uncommon to be confronted with a situation where it is either impossible or very highly impractical to study the entire data set. For example, if we are light bulb manufacturers and we want to learn how long our light bulbs last, we could ask every customer to track the lifetime of every bulb and report back to us, but we know that this is not going to happen. We could run all of our light bulbs ourselves until they burn out, but this is not a very good business plan. Instead, we take a representative sample of the bulbs we manufacture, run them ourselves until they burn out, and then study our data. In a later chapter I will discuss what it takes for a sample to be representative of the entire population. We need to know what fraction of the manufactured bulbs we must choose as samples, what rules we need to follow in selecting these samples, and just how well the results of studying the sample population predict the results for the entire population.

I hope that these examples and the explanations have created enough interest that you'd like to continue reading. The next two chapters will present most of the basic mathematics and definitions needed to continue. Then starting in Chapter 4 I will concentrate on examples of how random variables and probability affect many of the things we do and also how they are at the basis of many of the characteristics of the world we live in.

CHAPTER 2

PROBABILITY DISTRIBUTION FUNCTIONS AND SOME BASICS

This is the mathematics-background chapter. It can be skimmed lightly just for understanding the definitions of terms that will keep appearing in later chapters or it can be followed carefully. The mathematics in this chapter is almost entirely nothing more than combinations of adding, subtracting, multiplying, and dividing some numbers. Since in many cases I'll be dealing with very many numbers at a time (hundreds or even thousands), it is impractical to write out all of the numbers involved in a given calculation. What I do instead is introduce some fancy summarizing notation that includes things such as *subscripted variables, summation signs*, and so on. Don't let these words frighten you away—learning to work with them is no harder than adding a few new words to your vocabulary. You already know the mathematical operations involved, so this is like learning to cook a recipe from a new cookbook when you already are a pretty good cook.

THE PROBABILITY DISTRIBUTION FUNCTION

The Probability Distribution Function (PDF) is a powerful tool for studying and understanding probabilities in different situations. In order to talk about PDFs, I must first discuss the general idea of a mathematical function. A mathematical function—or, more simply, a function—is the mathematical equivalent of a food processor: You pour in one or more numbers, push the "grind" button for a few seconds, and pour out the resulting concoction. Three important facts just came past in this simple analogy:

1. One or more numbers, usually called *variables* or *independent variables* go into the function, depending upon the particular function "recipe" we're dealing with.
2. The function somehow *processes* these numbers to produce a result (a number) that is usually called the *value of the function* for the particular value of the variable(s) that went into the function.
3. The function produces exactly one value for any given input variable(s).

Item 1 above is true in the general case, but the cases we will be dealing with only have one number going into the function—so from now on we'll limit our discussion to this case. As an example, consider the function "Double the number coming in and then add 3 to it." If the number coming in is 5, the result is 13; if the number coming in is 2, the result is 7; and so on.

In simple mathematical notation, the input number is represented by a letter, often x, almost always letters from near the end of the alphabet. x can take on many possible values, hence the name "variable," and the above function *of the variable x* is written as

$$\text{function} = 2x + 3$$

Sometimes a function must be written as one or more rules. For example, the *absolute value* function of [the variable] x is defined by the following two rules:

1. If x is equal to or greater than 0, the function equals x.
2. If x is less than 0 (is negative), the function equals x with the minus sign taken away.

Examples of the absolute value function are

$$x = 2, \qquad \text{function} = 2$$
$$x = 4, \qquad \text{function} = 4$$
$$x = -3, \qquad \text{function} = 3$$
$$x = -6.4, \qquad \text{function} = 6.4$$
$$x = 0, \qquad \text{function} = 0$$

The absolute value function is a common enough function for a special notation to have been developed for it: Absolute value of x = |x|.

In both of the above examples, [the variable] x could have been any number that you can think of. Sometimes, however, the allowed values of x are restricted. For example, suppose our function is "the probability of getting a particular number that is the sum of the two faces of a pair of dice when we roll these dice." The only allowed numbers for the input variable are the possibilities that you can get when you roll a pair of dice—that is, the integers from 2 to 12. This function was described in the last chapter. If the input variable is 2, the function is 1/36; if the input variable is 3, the function is 2/36; and so on.

The class of functions that describe probabilities is known as probability distribution functions (PDFs). All PDFs share several attributes:

1. The function can never be less than 0. We have no concept of what a negative probability might mean.
2. The function can never be greater than 1. Remember that a probability of 1 describes a certain event. A probability greater than 1 just doesn't mean anything.
3. When the allowed inputs to a function can be written as a finite list of numbers, the sum of the values of the function for all of the allowed inputs must equal 1. This situation was also described in Chapter 1. Every allowed number going into the function has associated with it a result, or value of the function. Since the list of allowed numbers covers (as its name implies) every possible input number to the function, then all of the values of the function added together (the sum of all the possible values of the function) must be a certain event (i.e., *one* of these things *must* happen), and this sum must equal one.

The numbers going into a PDF are *variables* which are numbers chosen from an allowed set of numbers (the situation usually makes it very clear as to just what the allowed set of numbers is). When we invoke a particular situation, then one of the allowed variables is chosen, seemingly at random, according to the likelihoods assigned by the probabilities attached to all of these [allowed] random variables.

A very useful way of looking at PDFs is to really *look* at the PDF (no pun intended) by drawing a picture, or graph, of the function. We start with a horizontal line with evenly marked steps, just like a ruler. The numbers on this line will represent the allowed values of the input variable, x. Although it's not absolutely necessary, the numbers are usually spaced as if we were drawing a piece of a ruler. This line is sometimes called the "x axis," sometimes the "abscissa." For our pair of dice example, the x axis is shown in Figure 2.1.

The allowed values of x are the integers from 2 through 12. Figure 2.1 shows numbers from 0 to 15. So long as the allowed values of x are shown, it doesn't matter if some "extra" numbers are also shown. This is the choice of the

Figure 2.1. The x axis of the PDF graph for the roll of a pair of dice.

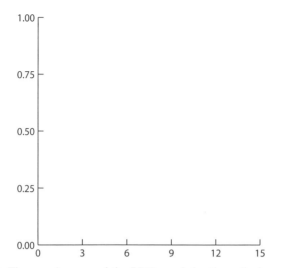

Figure 2.2. The x and y axes of the PDF graph for the roll of a pair of dice.

person drawing the graph, and this choice is usually determined by the drawer's opinion of what will make the clearest-to-understand graph. If for some reason not all of the allowed values are shown, then the graph is a graph of only part of the PDF and should be carefully labeled as such.

Next we add a vertical line with evenly marked steps (Figure 2.2). The numbers on this line represent the values of the function. Since this is a probability, we know that the only possible numbers are between 0 and 1.

We put a dot (or an x, or any convenient marking symbol) at the intersection of a vertical line drawn upward from a particular value of x and a horizontal line drawn across from the probability corresponding to that value of x. Our graph of the PDF for tossing a pair of dice, Figure 2.3, is now complete. The choice of showing x values from 0 to 15 was made to emphasize that there is 0 probability of x being less than 2 or greater than 12.

In this example the largest probability is less than 0.2. In order to make the graph easier to read, therefore, we can zoom in on the vertical axis. Figure 2.4 shows the same PDF as Figure 2.3, but with the vertical axis running from 0.0 to 0.2.

In this example, the PDF is symmetric about the point $x = 7$. This means that everything to the right of $x = 7$ is a mirror image of everything to the left of $x = 7$ (and of course vice versa). $x = 7$ is called the *axis of symmetry*. PDFs do not have to be symmetric about any value of x and we shall see many examples of unsymmetrical PDFs. Symmetric PDFs do, however, show up so

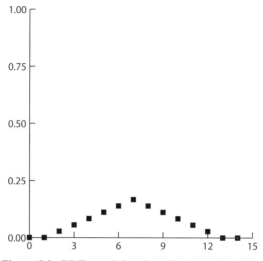

Figure 2.3. PDF graph for the roll of a pair of dice.

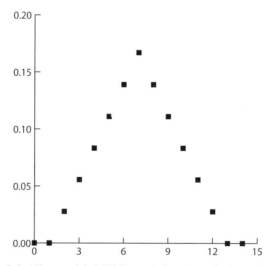

Figure 2.4. "Zoomed–in" PDF graph for the roll of a pair of dice.

often that some special rules about their properties will be developed in a few pages.

Sometimes I will be looking at a PDF that has hundreds or possibly thousands of allowed values of x. When this happens, I might choose to replace these hundreds of dots with a continuous line. This is just a convenience to make the graph easier to look at. This is a very important point. The PDF as I defined it is really known as a *discrete* PDF. This means that there is a finite number of allowed values of x, each represented by a point on the graph.

Drawing a line between these points (and then forgetting to draw the points) does *not* mean that we can allow values of x in between the x point locations. For example, a single die roll can yield values of x which are the integers from 1 to 6; 3.5 or 4.36 or any other non-integer value simply does not make sense. On the other hand, there is a category of PDFs known as *continuous* PDFs which does allow for any value of x—possibly between two limits or possible even *any* number x. An example of this is the probability of a certain outside temperature at noon of a given day (any value of x allowed, but between about $-60\,°F$ and $+125\,°F$, should cover most places where you'd consider living). Another example is the probability of the weight or height of the next person coming through the door. The same concepts of probability and PDF hold for continuous as for discrete PDFs, but we must reconsider what we mean when we say that the "sum of the probabilities must be one." If we are to allow any value of x between, say 1 and 2, then we must allow 1, 1.1, 1.11, 1.111, 1.1111, and so on. How can we add all of these probabilities up? The answer lies in the integral calculus—we learn to "add up" an infinite number of cases, each of which applies to an infinitesimally small region on the x axis. Don't worry about this unless you want to.

I could show some more PDF examples here, but the rest of the book will be treating so many examples of PDFs and their graphs that we might as well just look at them as they come along.

AVERAGES AND WEIGHTED AVERAGES

I'd like to step aside for a moment and discuss the idea of a *weighted average* and of a very powerful, concise, notation for discussing various calculations. I'll start with the calculation of a simple average of a few numbers. Consider the list of 10 numbers {1, 2, 3, 3, 3, 4, 5, 5, 5, 5}. The average of these numbers, also called the *arithmetic mean* or often simply the *mean*, is just the sum of these numbers divided by how many numbers we just summed:

$$\frac{1+2+3+3+4+5+5+5+5}{10} = \frac{36}{10} = 3.6$$

Now for the notation. Keep in mind that nothing new is happening here, I'm just writing things in a more concise "language" that suits our needs better than words.

I'll start with the idea of subscripted variables. In this example, there are 10 numbers in a list. The first number is 1, the second number is 2, the third number is 3, the fourth number is also 3, and so on. Generalize the idea of referring to a specific case of x by adding a subscript i: x_i now refers to the "ith" number in the list. Remember that the values of i have nothing to do with what the numbers in the list are, they just tell you where in the list to look for a number:

$$x_1 = 1$$

$$x_2 = 2$$

$$x_6 = 4$$

and so on

I say that I want to add up (get the sum of) this list of numbers by using the uppercase Greek letter sigma: Σ.

The way to write that I'm going to add up the list of numbers is

$$\text{sum} = \sum_{i=1}^{10} x_i$$

This is a handful for someone who hasn't seen it before. Taking it apart piece by piece, the uppercase sigma simply means that I'm going to be adding something up. What am I adding up? I'm adding up whatever appears to the right of the sigma, in this case a bunch of x_i's. How many of them am I adding up? Look under the sigma and you'll see $i = 1$. That means start with x_1. Look on top of the sigma and you'll see 10. This means $i = 10$ (it's not necessary to repeat the i). That means stop after x_{10}. Putting all of this together, we obtain

$$\text{sum} = \sum_{i=1}^{10} x_i = x_1 + x_2 + \cdots + x_{10} = 1 + 2 + 3 + 3 + 3 + 4 + 5 + 5 + 5 + 5 = 36$$

I snuck in another notational trick above, whose meaning I hope is obvious: . . . means "and all the items in between." To complete the picture of how to use this notation, suppose I only want to sum up the first four numbers of the list:

$$\text{sum} = \sum_{i=1}^{4} x_i = x_1 + x_2 + x_3 + x_4 = 1 + 2 + 3 + 3 = 9$$

One last comment: Occasionally you'll see a sigma with nothing above or below it. This is an informal notation that means "Look to the right, what we're adding up should be obvious. From the context of whatever you see, you should be able to identify everything that can be added up—go ahead and add them all up."

Armed with this notational tool, I can write a formula for the average of our set of numbers as

$$\text{average} = \frac{1}{10} \sum_{i=1}^{10} x_i$$

The general expression for the average of a list of n numbers is then

$$\text{average} = \frac{1}{n} \sum_{i=1}^{n} x_i$$

This formula is so general that it's useless standing there by itself. Someone has to provide you with the list of n numbers (or instructions how to create this list) before you can use it.

As an example, suppose $n = 3$. I could give you the list

$$x_1 = 9, \qquad x_2 = 16, \qquad x_3 = 25$$

from which you could calculate the average: $(9 + 16 + 25)/3 = 50/3 = 16.6667$.

Or, I could give you set instructions for creating the x_i:

$$x_i = (i + 2)^2$$

Or I could choose to write the instructions right into the formula:

$$\text{average} = \frac{1}{n} \sum_{i=1}^{n} (i + 2)^2$$

These are all absolutely equivalent ways of writing the same thing. The first way, listing the values of x_i, is fine when there are only a few values. What if $n = 100,000$? You wouldn't want to list them all. On the other hand, if the list of numbers is the result of data taking (for example, the number of buttons on the shirts of the first 2000 people you saw on the street this year), then there is no set of instructions and the list is the only choice.

The choice between the second and the third way is usually just a stylistic preference of the author of whatever you're reading. Either of these is preferable to a list when the set of instructions (aka the function) exists because it lends more insight into how the list of numbers came about than does just showing the list.

Returning to the example above, I'll repeat the list of 10 numbers so that we have it in front of us: {1, 2, 3, 3, 3, 4, 5, 5, 5, 5}. The number 3 appears 3 times and the number 5 appears 4 times. Another way of specifying this list is by writing a shorter list of numbers showing each number that appears only once, namely {1, 2, 3, 4, 5}, and then creating a second list called a *Weighting Function*, which I'll represent with the letter w_i. The weighting function tells us how many times each of the numbers in our main list appears. In this case, the weighting function is {1, 1, 3, 1, 4}. Putting these together, these lists tell us that 1 appears once, 2 appears once, 3 appears 3 times, 4 appears once, and 5 appears 4 times.

In our shorthand notation, the sum of the original 10 numbers is now

$$\text{sum} = \sum_{i=1}^{5} w_i x_i = (1)(1) + (1)(2) + (3)(3) + \cdots + (4)(5) = 36$$

and the average is

$$\text{average} = \frac{\sum_{i=1}^{5} w_i x_i}{\sum_{i=1}^{5} w_i}$$

Where did the summation in the denominator come from? Look back at the definition of the weighting list. The ith item in the weighting list tells us the number of times we will see w_i in the main function list. If we add up the numbers in the weighting list, we must get the total number of items in the list. Going back to the earlier definition of average, this is just what I called "n", the number of items in the list.

EXPECTED VALUES

Suppose the weighting function is the PDF that is associated with the variables x_i? I'll explain why I want to do this soon, for now let's just go through the calculations. In terms of notation, just to keep it clear that I want the weighting function to be a PDF, I'll replace w_i with p_i in the weighted average expression:

$$\text{average} = \frac{\sum_{i=1}^{n} p_i x_i}{\sum_{i=1}^{n} p_i}$$

As was discussed above, the sum of all the possible probabilities must equal 1. The summation in the denominator is therefore 1, and we don't have to carry it around any more:

$$\text{expected value} = \sum_{i=1}^{n} p_i x_i$$

This weighted average is given the name *expected* value of the random variable x. Remember that the weighted average as described above is just an average of a list of numbers with each number weighted by the number of times that number occurs when calculating the average. When the weighting function is a probability, we're generalizing "how many times that number occurs" to "what's the likelihood of that number occurring each time we try this?"

I suspect that some examples are in order. Let x_i represent the number of hours that a certain brand and model light bulb will last. The manufacturer has measured thousands of these bulbs and has characterized their lifetime to the nearest 100 hours. Table 2.1 shows the probability of bulb lifetimes.

Just as a check, add up all the probabilities and you'll get 1.00. If you had gotten other than 1.00, you'd know that there was a mistake made somewhere. The graph of this PDF is shown in Figure 2.5.

Let's write a few cases in terms of the subscripted variable notation: $x_4 = 1100$, $x_8 = 1500$, $p_4 = .22$, $p_9 = .01$, and so on.

The expected value of x is therefore

TABLE 2.1. Light Bulb Manufacturer's PDF

i	Lifetime	Probability
1	800	.02
2	900	.04
3	1000	.10
4	1100	.22
5	1200	.30
6	1300	.21
7	1400	.07
8	1500	.03
9	1600	.01

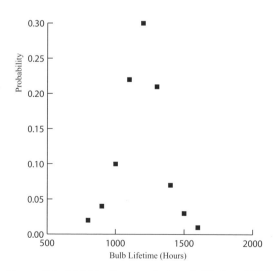

Figure 2.5. Light bulb manufacturer's lifetime PDF.

$$\text{expected value} = \sum_{i=1}^{9} x_i p_i = (800)(.02) + (900)(.04) + (1000)(.10) + (1100)(.22)$$
$$+ (1200)(.30) + (1300)(.21) + (1400)(.07)$$
$$+ (1500)(.03) + (1600)(.01) = 1186$$

This means that if you were to buy a few thousand of these bulbs, you should expect the average lifetime to be about 1186 hours.

THE BASIC COIN FLIP GAME

In Chapter 8 I will spend a lot of time talking about a simple coin flip gambling game. I'd like to introduce this game here; it's a simple game to understand, and it illustrates many of the characteristics of the expected value. The rules of the game are simple: One of us tosses (flips) a coin. If the coin lands heads up, you give me a dollar, if it lands tails up, I give you a dollar. Assuming a fair coin, the probability of either occurrence is 0.5. Taking the winning of a dollar as a variable value of +1 and taking the losing of a dollar as the value of −1, the expected value is simply

$$\text{expected value} = \sum_{i=1}^{2} x_i p_i = (+1)(.50) + (-1)(.50) = .50 - .50 = 0$$

This seems odd: If we flip a coin just once, I must either win a dollar or lose a dollar. There are simply no other choices. We seem to have a contradiction here: In this game I can never *expect* to actually see the *expected* value. Actually, what the expected value is telling us is that if we flip the coin thousands of time, I should expect the difference between my winnings and my losses to be only a small percentage of the number of times a dollar changed hands.

I'd like to return for a moment to the topic of a symmetric PDF. Figure 2.6 is an example of a symmetric PDF. A vertical line drawn through $x = 3$ is the axis of symmetry, with 3 being the value of x from which things look the same both to the right and to the left.

The expected value of this PDF, calculated in detail, is

$$E(x) = (.1)(1) + (.2)(2) + (.4)(3) + (.2)(4) + (.1)(5) = 3$$

The expected value of x is exactly the value of the axis of symmetry. Is this a coincidence, or will this always be the case? I'll explore this question by rewriting the expected value calculation in a particular manner: For each value of x, I'll write x in terms of its distance from the axis of symmetry. That is, when $x = 1$ I'll write $x = 3 - 2$; when $x = 4$ we'll write $x = 3 + 1$, and so on. The expected value calculation is now

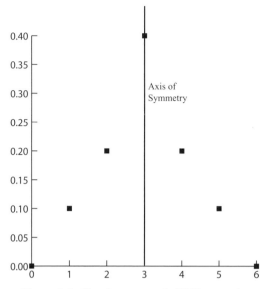

Figure 2.6. Simple symmetric PDF example.

$$E(x) = (.1)(3-2)+(.2)(3-1)+(.4)(3)+(.2)(3+1)+(.1)(3+2) = 3$$

Rearranging the terms a bit, we obtain

$$E(x) = (.1)(3-2)+(.1)(3+2)+(.2)(3-1)+(.2)(3+1)+(.4)(3) = 3$$

For the term containing $(3 - 2)$, there is an identical term containing $(3 + 2)$. For the term containing $(3 - 1)$, there is an identical term containing $(3 + 1)$. The symmetry property guarantees that there will be a balance of these terms for every term except the term $x = 3$ itself.

Now I'll do some more manipulation. Just look at the term containing $(3 - 2)$ and the term containing $(3 + 2)$:

$$(.1)(3-2)+(.1)(3+2) = (.1)(3)-(.1)(2)+(.1)(3)+(.1)(2) = (.1)(3)+(.1)(3)$$

What has happened? The $-(.1)(2)$ and $+(.1)(2)$ pieces have canceled out, as if they had never been there in the first place. This operation can be repeated for every pair of terms present, leaving us with

$$E(x) = (.1)(3)+(.2)(3)+(.4)(3)+(.2)(3)+(.1)(3)$$
$$= (.1+.2+.4+.2+.1)(3) = (1)(3) = 3$$

and since we know that this is not an accident and that the probabilities must *always* add up to 1, we see that for a symmetric PDF the expected value of x must always be the axis of symmetry itself.

This example of a symmetric PDF has an odd number of x values. What about a symmetric PDF with an even number of x values? With a bit of thought and perhaps some sketching on a napkin, you should be able to convince yourself that a symmetric PDF with an even number of x values cannot have an x value at the axis of symmetry (if it did, then there would have to be one x value more to the left than to the right, or vice versa).

In the case of a symmetric PDF with an odd number of points, if the point on the axis of symmetry also is the point that has the largest probability, then the expected value of x is also the most likely value of x.

Another special case is that of a PDF where the probabilities of all the variable values are the same. This may be a symmetric PDF, but it doesn't have to be symmetric—as Figure 2.7 shows. The variable (x) values are 0, 1, 2, 4, and 7.

If all of the probabilities are the same and they all must add up to 1, then these probability values must of course be = 1/(number of values). In this case, there are 5 random variable values, so each probability must be just $1/5 = 0.2$.

The expected value of x is

$$E(x) = (.2)(0) + (.2)(1) + (.2)(2) + (.2)(4) + (.2)(7)$$

Rearranging this by "pulling out" the common factor of .2, we obtain

$$E(x) = (.2)(0 + 1 + 2 + 4 + 7) = \frac{0 + 1 + 2 + 4 + 7}{5}$$

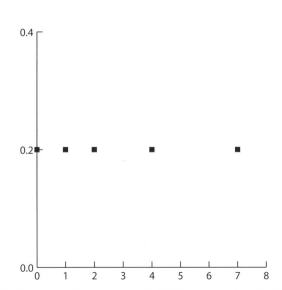

Figure 2.7. Example of an asymmetric PDF with equal probability values.

When all of the probabilities are equal, the expected value is just the average value of the random variable values.

A word about averages: The average as calculated above is sometimes called the *arithmetic average*, to distinguish it from the *geometric average*, which I won't worry about here. It is also sometimes called the *arithmetic mean*. The mean must be no smaller than the smallest value in the group and no larger than the largest value in the group. Intuitively, it's a number that somehow "summarizes" a characteristic of the group. For example, if I told you that the mean age of a class of children is 7.2 years old, you would guess that this might be a second-grade class, but it is certainly not a high school class. Another indicator about the properties of a group of numbers is the *median*. The median is the number that has as many numbers to the right of it (on the graph) as it does to the left of it. In the case of a symmetric PDF with an odd number of terms, the median x value is identically the mean as well as the axis of symmetry. A third measurement is called the *mode*. The mode is the number in a list of numbers that is repeated the most. In the list of numbers used a few pages ago to explain indices and weighted averages, the number 5 is the mode.

Knowing, or knowing how to calculate, the expected value of a PDF tells us something about "what to expect" from the experiment whose random variable values and their probabilities made up our PDF. It does not, however, tell us as much as there is to know.

THE STANDARD DEVIATION

Consider the following example: You are a manufacturer of glass marbles. You have a very tightly controlled process that manufactures 1-oz marbles. You sell these marbles in 3-lb (48-oz) bags, so there should be 48 marbles in each bag. For the past few months, you've been having trouble with the machine that loads the marbles into the bags—it's gotten erratic. Just to be safe until you get the machine fixed, you decide to put (set your machine to put) 50 marbles into each bag so that no customer ever gets less than 48 marbles—but still, there are customer complaints.

The easiest way to track how many marbles are loaded into each bag is to count some of them. You grab 100 bags of marbles and count the number of marbles in each bag. Table 2.2 shows what you find.

No wonder some customers are complaining! While most of the customers are getting 48 or more marbles, some customers are getting fewer marbles than they've been promised!

Table 2.2 is easily converted into a PDF. If there were 100 bags of marbles weighed, then the probability of getting 50 marbles is just 35/100 = .35, and so on. In other words, if we take the numbers on the vertical (or "y") axis and divide them by the total number of points (bags of marbles) in our sample, then we have a proper PDF.

TABLE 2.2. Marble Manufacturer's 100-Bag Production Sample

Number of Marbles in Bag	Number of Bags	Number of Marbles in Bag	Number of Bags
42	2	49	20
43	0	50	35
44	0	51	14
45	0	52	0
46	0	53	14
47	8	54	0
48	3	55	4

To really know exactly what your marble factory is shipping, you would have to count the number of marbles in each and every bag of marbles that you produce. Since this is extremely time-consuming, you would like to *sample* your production by only selecting a bag of marbles to count every now and then. The question is, What do I mean by "now and then" and how do you know when your sample population truly represents your entire distribution? One extreme is counting the marbles in every bag. Alternatively, you could count the marbles in a bag pulled off the shipping dock once a month. Or, you could grab a bag every hour until you have 100 bags and look at the distribution. While the first two choices are not very good (too time-consuming or not representative of the true distribution), how do we know just how useful the information we're getting from the last choice really is? I'll leave this discussion for another chapter and just assume for now that the 100-bag sample represents the factory production very well.

Figure 2.8 clearly shows that the bag-filling machine is not doing a good enough job of counting marbles. Let's assume that you reworked the machine a bit, ran it for a while, repeated the 100-bag counting exercise, and got Figure 2.9. This PDF indicates that fewer customers are now getting less than 48 marbles, but the situation still isn't great. Let's refer to the PDFs of Figures 2.8 and 2.9 as *Poor* and *Marginal*, respectively, and tabulate them in Table 2.3.

Calculating the expected value of both distributions, without showing the arithmetic here, I get 50.10 for the *Poor* distribution and 49.97 for the *Marginal* distribution. This is odd. My goal is to minimize (ideally, to eliminate) bags with less than 48 marbles. It would seem as if raising the expected value of the number of marbles in the bag would make things better—but in this case the distribution with the higher expected value gives worse results. In order to measure just how good the marble bag machine is (or isn't) working, determine how to specify an operating tolerance, and perhaps set warning levels for "adjusting the machine," and "shutting down production," I need a way to quantify the difference between these two distributions in a meaningful

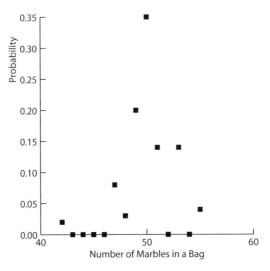

Figure 2.8. PDF of marble bag count.

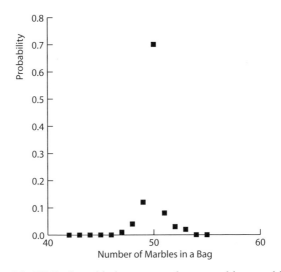

Figure 2.9. PDF of marble bag count after reworking machinery.

way. Clearly, there is something more going on here than the expected value alone is telling us.

I am going to rewrite Table 2.3 a bit. For each distribution, I'll show the number of marbles in the bag in terms of their difference from the expected value (Table 2.4).

The numbers in the two new columns in Table 2.4 are both positive and negative, indicating values higher than or lower than the expected value(s),

TABLE 2.3. Marble Manufacturer's *Poor* and *Marginal* PDFs

Number of Marbles in Bag	"Poor" Probability	"Marginal" Probability	Number of Marbles in Bag	"Poor" Probability	"Marginal" Probability
42	0.02	0	49	0.20	0.12
43	0	0	50	0.35	0.70
44	0	0	51	0.14	0.08
45	0	0	52	0	0.03
46	0	0	53	0.14	0.02
47	0.08	0.01	54	0	0
48	0.03	0.04	55	0.04	0

TABLE 2.4. Table 2.3 Rewritten to Show Differences from Expected Values

Number of Marbles in Bag (x)	Poor		Marginal	
	Probability	Number − Expected Value	Probability	Number − Expected Value
42	.02	42 − 50.10 = −8.10	0	42 − 49.97 = −7.97
43	0	43.00 − 50.10 = −7.10	0	−6.97
44	0	−6.10	0	−5.97
45	0	−5.10	0	−4.97
46	0	−4.10	0	−3.97
47	.08	−3.10	.01	−2.97
48	.03	−2.10	.04	−1.97
49	.20	−1.10	.12	−0.97
50	.35	−.10	.70	+0.03
51	.14	+0.90	.08	+1.03
52	0	+1.90	.03	+2.03
53	.14	+2.90	.02	+3.03
54	0	+3.90	0	+4.03
55	.04	+4.90	0	+5.03

respectively. We aren't as much interested in whether these numbers are higher or lower than the expected value(s) as we are in how far away they are from their respective expected values. I can keep the information we're interested in while eliminating the information we're not interested in by simply squaring these numbers. Squaring means multiplying a number by itself. Remember that a negative number times another negative number is a positive number; for example, $(-4)(-4) = (4)(4) = 16$. The square of all of our numbers must therefore be positive numbers. Table 2.5 shows the results of squaring columns 3 and 5 of Table 2.4.

Multiplying the entries in columns 2 and 3 and then summing these products gives us the expected value of the squared error numbers for the *poor*

TABLE 2.5. Table 2.4 Rewritten Using Square of Differences from Expected Values

Number of Marbles in Bag (x)	Poor		Marginal	
	Probability	(Number – Expected Value)^2	Probability	(Number – Expected Value)^2
42	.02	65.61	0	63.82
43	0	50.41	0	48.58
44	0	37.21	0	35.64
45	0	26.01	0	24.70
46	0	16.81	0	15.76
47	.08	9.61	.01	8.82
48	.03	4.41	.04	3.88
49	.20	1/21	.12	0.94
50	.35	0.01	.70	0.00
51	.14	0.81	.08	1.06
52	0	3.61	.03	4.12
53	.14	8.41	.02	9.18
54	0	15.21	0	16.24
55	.04	24.01	0	25.30

PDF. Similarly, using columns 4 and 5 gives us the expected value for the squared errors for the marginal PDF. The results of these calculations are called the *variance*s of these PDFs, 4.71 for the poor PDF and 0.75 for the marginal PDF. Variance is a measure of how much we can expect results to *vary* from their expected value with repeated trials. While the marginal PDF has the higher expected value of these two PDFs, which we might think means that we're going to get many unhappy customers, the marginal PDF's variance is so much the lower of the two that this PDF would actually result in fewer customers getting fewer marbles than the package promises. Another way of putting it is that the lower the variance, the more tightly the numbers will be "bunched" about the expected value. An interesting property of squaring the error terms is that since the squares of numbers climb much faster than do the numbers themselves ($2^2 = 4$, $3^2 = 9$, $4^2 = 16$, etc.), one moderate deviation will have a much larger effect on the variance than will several small deviations.

This property of the variance has made it almost the universal approach to characterizing how well a list of numbers is grouped about its mean. While I described the variance for a PDF, the same definition can be used for any list of numbers. Using the tools of the calculus, the same definition can be extended to continuous PDFs.

An annoying factor when dealing with variances is that they do not have the same *units* as the underlying random variable. In the above example, we are dealing with numbers of marbles, but the variance is measured in (number of marbles)^2. In order to be able to picture what to expect, it is common to use the square-root of the variance, known as the *standard deviation*. In this example, the standard deviation for the poor PDF is 2.2, while the standard deviation for the marginal PDF is 0.7. Standard deviation is used so often in probability and statistics calculations that its usual symbol, σ (lowercase Greek sigma), is universally recognized as standing for the standard deviation and often people will simple say "sigma" when they mean "standard deviation."

Next I'll consider another example of a PDF, its expected value, and its standard deviation. The simplest distribution to picture is the *uniform* distribution. A uniform distribution of *n* events is defined as a distribution in which each event is equally likely. The probability of each event is therefore simply $1/n$. (The coin flip PDF is an $n = 2$ uniform distribution in which each event has a probability of $1/n = 1/2$.)

Figure 2.10 shows the PDF of a uniform distribution of integers. From the figure you can see that a uniform PDF has a starting, or lowest, value of *x* (1 in this example) and a stopping, or highest, value of *x* (20 in this example). The expected value of this distribution is very easy to calculate because the probabilities are all the same so they don't really need an index (the subscript *i*), and the probability can be written outside the summation sign:

$$\text{expected value} = \sum_{i=1}^{n} p_i x_i = p \sum_{i=1}^{n} x_i = \frac{1}{n} \sum_{i=1}^{n} x_i$$

Come to think of it, this is exactly the formula for the simple average of *n* items x_i, which is a property of uniform distributions that was described earlier in this chapter. To formally evaluate this expression, I would need to figure out how to express the locations of each of the x_i depending on the lowest value, the highest value, and the number of points. This isn't really that big a

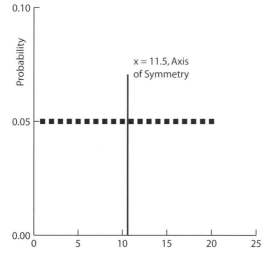

Figure 2.10. PDF of a uniform distribution of integers.

job, but fortunately it's an unnecessary job. Just look at the graph of this PDF—regardless of the lowest value, the highest value, and the number of points, this graph is symmetric about some value of x. Let's call the lowest x value of x_{low} and the highest value x_{high} (not very imaginative choices, but they'll do the job). I want to find the value of x that is equidistant from x_{low} and x_{high}, or, in other words, the midpoint of the line between them. I know that this point will be the expected value of x for a symmetric PDF.

$$\text{expected value} = \frac{x_{low} + x_{high}}{2}$$

For the PDF shown in Figure 2.10, the expected value of x is therefore 11.5. Interestingly enough, in calculating this expected value, there is no mention of how many values of x_i, (n), there are in the distribution.

Now I'll look at some standard deviations of a uniform distribution between $x = 1$ and $x = 3$. For a 3-point distribution between $x = 1$ and $x = 3$, I get (Table 2.6) an expected value (E) of 2 and a standard deviation of 0.816.

Next I'll move this same distribution over a bit so that I have a three-point uniform distribution between $n = 0$ and $n = 2$, as shown in Table 2.7.

Comparing the above two distributions, we see that the expected values are different. That this should happen is clear from the symmetry argument—the expected value has to be in the "middle" of the distribution. The standard deviation, on the other hand, is the same for both distributions. This is because

TABLE 2.6. Three-Point Uniform Distribution Between $x = 1$ and $x = 3$

i	x_i	P_i	$x_i P_i$	$(x_i - E)^2$	$P(x_i - E)^2$
1	1	0.333	0.333	1	0.333
2	2	0.333	0.667	0	0
3	3	0.333	1.00	1	0.333
			$E = 2$		Variance = .667
					$\sigma = .816$

TABLE 2.7. Three-Point Uniform Distribution Between $x = 0$ and $x = 2$

i	x_i	P_i	$x_i P_i$	$(x_i - E)^2$	$P(x_i - E)^2$
1	0	0.333	0.00	1	0.333
2	1	0.333	0.333	0	0
3	2	0.333	0.667	1	0.333
			$E = 1$		Variance = .667
					$\sigma = .816$

standard deviation is a property of the shape of the distribution, not its location on the x axis.

Now look at two other cases. I'll narrow the distribution so that it only extends from 0 to 1 and I'll widen it so that it extends from 0 to 4. The calculations parallel those above, so I'll skip showing the details and just show the results in Table 2.8. For a three-point uniform distribution of width X_{high} − X_{low}, it appears that the standard deviation (sigma) is scaling directly with the width of the distribution. In other words, the standard deviation is a measure of how much the points (in this example, three points) are spread out about the expected value. If all three points lied on top of each other (e.g., they were all exactly at $x = 2$), then the standard deviation would be identically 0.

Let's try changing some other parameters. I'll keep the width of the distribution constant but vary the number of points. Returning to the distribution between 1 and 3, Table 2.9 shows σ for several numbers of points (n).

While the expected value of x does not depend on the number of points in the uniform PDF, the standard deviation clearly does. Figure 2.11 shows how the standard deviation varies with the number of points. A calculation that is outside of the scope of this book shows that, as n gets very large (approaches infinity), the standard deviation will approach 0.577. At $n = 100$ the standard deviation is .583, which is only about 1% bigger than the standard deviation for an infinitely large number of points. In general, for n very large, the standard deviation of a uniform distribution is

$$\sigma = \frac{X_{high} - X_{low}}{\sqrt{12}} = \frac{X_{high} - X_{low}}{3.464}$$

TABLE 2.8. Three-Point Uniform Distributions of Varying Widths

Width	Sigma
1	0.408
2	0.816
4	1.63

TABLE 2.9. Sigma of Uniform Distribution of Width 2 Versus Number of Points

n	σ
2	1
3	0.816
4	0.745

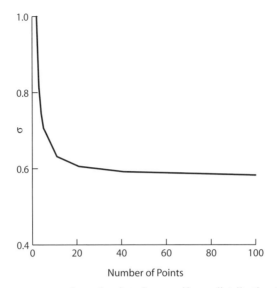

Figure 2.11. σ versus number of points for a uniform distribution 2 units wide.

THE CUMULATIVE DISTRIBUTION FUNCTION

I've shown how σ describes the "width" of a distribution about its expected value. Now let's see if I can quantify this. It will help to step aside for a moment and describe a new graph—the cumulative distribution function (CDF).

Return to the uniform PDF between $x = 1$ and $x = 3$. Consider a distribution with 21 points (Figure 2.12). The cumulative distribution is plotted by starting at an x value lower than any x value in the PDF and then stepping to the right (increasing x) at the same x step sizes as the PDF and plotting the sum of all the probabilities that have been encountered so far.

This isn't as difficult as it sounds. For $x < 1$ (x less than 1), the value of the CDF is 0. At $x = 1$, the PDF value is .048, and the CDF value is .048. For $x = 1.1$, the PDF value is .048, so the CDF value is $.048 + .048 = .096$. For $x = .2$, the PDF value is .048, so the CDF value is $.096 + .048 = .144$, and so on. This CDF is shown in Figure 2.13. I can immediately write down three important properties of the CDF:

1. If the PDF has a value of x below which the probability is 0, then the CDF is 0 below this same value of x. For a discrete PDF, this will always happen.
2. As x increases, the CDF can either increase or remain constant, but it can never decrease. This property comes straight out of the rule above for generating a CDF—the word "*subtract*" is never mentioned.

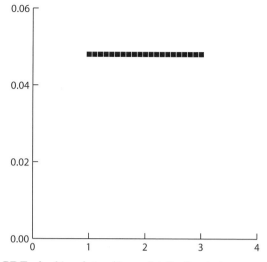

Figure 2.12. PDF of a 21-point uniform distribution between $x = 1$ and $x = 3$.

3. If the PDF has a value of x above which the probability is 0 (there are no more data points), then the CDF is 1 above this same value of x. This is because if we're above the value of x where all nonzero probabilities have been shown, then we must have covered every possible case and we know that all of the probabilities in a PDF must add up to 1.

From Figure 2.11 we see that this PDF has a σ of 0.6. Let's add this to and subtract this from the mean value of x, giving us 2.6 and 1.4. In Figure 2.13 there is a grey box whose left and right sides lie on 1.4 and 2.6, respectively. All of the points inside the box are the points that lie within 1 σ of the expected value of x. There are 13 points inside the box out of the total of 21 x values of the PDF. This means that 13 out of 21, or 62%, of the points lie within 1 σ of the expected value of x.

THE CONFIDENCE INTERVAL

Now let's take this from a graphics exercise back to a probability discussion. If I had a 21-sided die with the faces marked with the 21 values of this PDF and I rolled this die many, many times, I would expect the results to fall within 1 σ of the expected value of x 62% of the time. Since this is a symmetric PDF, I would also expect to fall below this range 19% of the time and above this range 19% of the time. For this distribution, we say that our 62% confidence interval is ± 1 σ wide.

The confidence interval is a measure of how well we *really* know a number. In our marble bag example the manufacturer must realize that his production

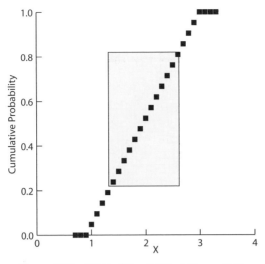

Figure 2.13. CDF of the PDF of Figure 2.12.

equipment can never be absolutely perfect. He can set an acceptable level of defective bags of marbles such as "1 in 100,000" and then, knowing the PDF, can calculate a σ that his machinery must at least meet so as to assure that, over the course of filling thousands of bags of marbles, he will never have to replace more than 1 bag in 100,000. The catch in the last sentence is *knowing the PDF.* The next chapter will show how, in most situations, this takes care of itself.

There are two issues that should be dealt with before this chapter can be considered finished. You really don't need to deal with either of them to understand the rest of the book. If you're not interested, take a peek and then either keep reading or just skip to Chapter 3.

FINAL POINTS

1. In the CDF discussion above, a 21-point PDF was used as an example. In the CDF the steps in x were 0.10 and the steps in y were about 0.05. We can talk about a region 1.00 wide, or 1.05 wide, or 1.10 wide, and so on, but we have to think about whether or not it makes sense to talk about a region, say, 1.12 wide. At least one of the end points of this region would fall in between x values and not be well-defined. If this PDF really came about from a 21-faced die, then these integer x points are all there are and we simply have to be careful what regions we try to talk about.

On the other hand, suppose we are weighing, say, fish from a commercial catch using a scale that only gives us weights to the nearest pound. We will end up with data that can be graphed as a PDF, but we know that the actual fish weights mostly fall between the points on the graph. In this example, our *measurement resolution* is 1 pound. Our "fish PDF" is just a discrete PDF approximation to what's really a continuous PDF. We should treat this mathematically by drawing a continuous PDF that in some sense is a *best fit* to the discrete data. How this is done in general is a college math course all by itself, but the next chapter will show how most practical cases turn out.

2. The definition of standard deviation given above gives an answer (zero) for a distribution that consists of only one point. This might not seem to be more than a nuisance because nobody would ever bother to calculate the standard deviation in this situation, much less ponder what it might mean. It does, however, reveal an underlying issue. Since we need at least two points to meaningfully discuss deviations of any sort, we really should base our standard deviation calculations on one less x variable point than we actually have. Probability texts refer to this as a "number of degrees of freedom" issue—we have one less degree of freedom than we have x variable points.

In order to correct for this error we go back to the definition of variance. The variance, if you recall, is the weighted average of the squares of the difference between the x points and the expected value. Since an average is basically a sum divided by the number of points being summed, n, we should replace n by $n - 1$ in the calculation. One way of doing this is to multiply the variance as defined by the term $n/(n-1)$. Since we're usually dealing with standard deviations rather than variances, we can apply this correction directly to the standard deviation by multiplying the standard deviation as defined by the square root of the above correction—that is, by $\sqrt{n/(n-1)}$.

If you've been trying to recreate the calculations in this chapter using computer programs such as a spreadsheet or a pocket calculator and are wondering why the standard deviations didn't agree, this correction is the answer to your problem.

Note that as n gets larger, this correction gets less and less important. For a small data set, such as $n = 3$, the correction is about 23%. For $n = 100$ the correction is about 0.5%, and for $n = 1000$ it's about .05%. This because $n/(n - 1)$ gets very close to 1 as n gets large, so multiplying by it gets less and less significant.

3. One last word about distinguishing between discrete and continuous PDFs. Figure 2.14a shows the PDF of a discrete uniform PDF consisting of 6 points between $x = 1.0$ and $x = 1.5$. Since the 6 probabilities must add up to 1 and they're all the same, each of them must be $= 1/6 = 0.167$. Figure 2.14b shows the CDF corresponding to this PDF. As with all CDFs, the value of the graph starts at 0 (to the left of any nonzero probabilities) and ultimately reaches 1 as x increases. If there were 60 points instead of 6, then each probability would equal 1/60. The CDF would be very busy, but would still go from 0 on the left to 1 on the right. If there were 600 points, each probability would be 1/600, and so on. With this many points I would probably choose to show the CDF as a continuous line rather than a large number of points, just for

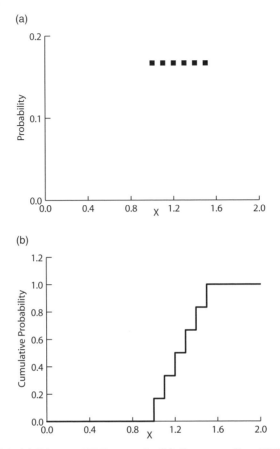

Figure 2.14. (a) Discrete PDF example. (b) Corresponding CDF example.

clarity. I would, however, be very careful to indicate that this is a discrete CDF.

Figure 2.15a shows a continuous uniform PDF with x going from 1.0 to 1.5, and Figure 2.14b shows its corresponding CDF. The CDF now can be interpreted as showing the "included probability" in the range from minus infinity to x. In other words, the probability of x being between 1.0 and 1.25 is 0.5, and so on. The CDF still starts at 0 on the left and ultimately reaches 1 as x increases. In some cases it's possible for the CDF to just keep getting closer and closer to 1 as x increases (and to 0 as x decreases). In the case of Figure 2.15, all non-0 probability is included between $x = 0$ and $x = 1.5$, so the CDF is 0 for all values of $x \leq 1.0$ and the CDF is 1 for all values of $x \geq 1.5$. The CDF represents the area included under the PDF between $x = 1$ and any value of $x > 1$. This means that the area under the total "box" that the PDF looks like must be 1.0. Since the width of the box is 0.5, the height of the box must be 2.0. This does not mean that the probability of x being, say, 1.1, is 2.0. 2.0 is

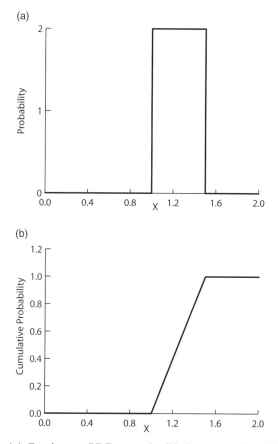

Figure 2.15. (a) Continuous PDF example. (b) Corresponding CDF example.

just the necessary height for this box if the area is to be 1, which in turn means that the probability of x being somewhere between 1.0 and 1.5 is the certain event, that is, 1.0.

Bottom line: PDFs for continuous functions and PDFs for discrete functions are somewhat different beasts. Drawing a PDF for a discrete function with a continuous line is inherently incorrect—I just do it for clarity sometimes. I'll clearly label what I'm doing. Theoreticians do handle situations of "an infinite number of infinitely close points" and similar puzzling issues, but this is far beyond the scope of this book (and, frankly, of my understanding).

CHAPTER 3

BUILDING A BELL

We need to understand what happens when there are many individual PDFs contributing to the final result in some situation. Rolling dice is an excellent example of this phenomenon. I'll admit that I've never studied the physics of rolling dice. I only know that there are six possible results for 1 die: the integers from 1 to 6. I guess landing on an edge and staying on that edge isn't impossible, but I'm not going to worry about it here. My understanding of the physics doesn't go much deeper than the observation that since all 6 possible results are equally likely, the PDF of the rolling a die is a uniform distribution.

I'll begin by examining in detail what happens when we add up the results of different random event experiments. In this case I will roll several dice or equivalently, one die several times. Understanding the concept of the probability distribution function (PDF) in Chapter 2 is central to this discussion; so if you skipped that chapter to get here, you might want to go back and read it before proceeding.

When a standard die is rolled, it lands with an integer between 1 and 6 facing up, every integer in this range being equally likely. Each integer has a probability of 1/6 (~.17) of occurring. The PDF is simple, as shown in Figure 3.1.

Now let's get a second die, identical to the first. I roll them both and am interested only in the sum of the two individual results. Clearly, this sum must

Probably Not: Future Prediction Using Probability and Statistical Inference,
by Lawrence N. Dworsky.
Copyright © 2008 John Wiley & Sons, Inc.

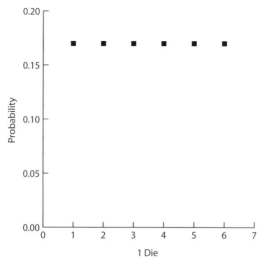

Figure 3.1. Rolling 1 die: A uniform 6-point PDF.

TABLE 3.1. Data for the PDF of Figure 3.2, Rolling 2 Dice

Sum	Combinations	Number of Ways to Get There	Probability
2	{1,1}	1	0.028
3	{1,2} {2,1}	2	0.056
4	{1,3} {2,2} {3,1}	3	0.083
5	{1,4} {2,3} {3,2} {4,1}	4	0.111
6	{1,5} {2,4} {3,3} {4,2} {5,1}	5	0.139
7	{1,6} {2,5} {3,4} {4,3} {5,2} {6,1}	6	0.167
8	{2,6} {3,5} {4,4} {5,3} {6,3}	5	0.139
9	{3,6} {4,5} {5,4} {6,3}	4	0.111
10	{4,6} {5,5} {6,5}	3	0.083
11	{5,6} {6,5}	2	0.056
12	{6,6}	1	0.028

be a number between 2 and 12 (1 on both dice and 6 on both dice, respectively). Table 3.1 shows all the possible results.

There are of course 36 possible combinations of the two dice, each combination having a probability of 1/36. However, since the sum must be an integer between 2 and 12, there are only 11 possible results. Also, except for sums of 2 and 12, there are more ways that one to get any result. Table 3.1 is organized to emphasize that I only care about the sums and the number of ways of getting to each sum.

Figure 3.2 is the PDF corresponding to Table 3.1, for the sum of two rolled dice. This PDF doesn't look at all like the PDF for one die. Actually, it's triangular.

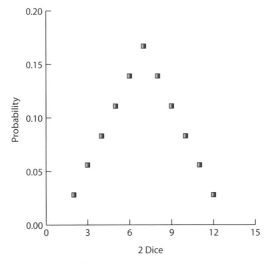

Figure 3.2. Rolling 2 dice: A triangular PDF.

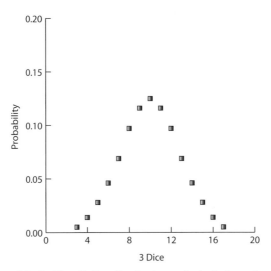

Figure 3.3. Rolling 3 dice: Beginnings of a bell-shaped curve.

I'll continue this exercise for a while, adding the results of summing more and more dice. I won't show all of the combinations because the tables get very large.

For 3 dice there are $6^3 = 216$ entries, with sums between 3 and 18. The shape of the distribution is still fairly triangular (Figure 3.3), but curves are appearing. From a distance, you might even say that this function is "bell-like."

Continuing, for the sum of 5 rolled dice I get a PDF that very definitely looks like a bell (Figure 3.4).

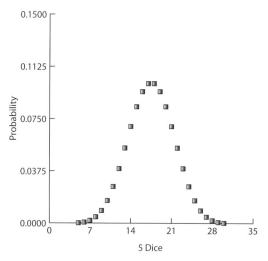

Figure 3.4. Rolling 5 dice: A bell-shaped curve.

Without showing more examples, I'll put forth the assertion that the sum of many rolls of a die produces a definitely bell-shaped PDF.

Since the PDF for one rolled die is simply a uniform distribution of integers (between 1 and 6), it's an easy leap to see that the sum of many uniform PDFs, regardless of their upper and lower limits, will always produce a bell-shaped curve as its PDF. Only the location of the peak (center of the curve) and the width of the curve will depend on the actual upper and lower limits of the original PDFs.

At this point it merely looks like an interesting curiosity has been demonstrated; the sum of many uniform PDFs produces a bell-shaped PDF. Actually, however, several other things have been shown. Remember that the sum of just two uniform PDFs produces a triangular PDF. This means that the PDF for, say, the sum of ten uniform PDFs is also the PDF for the sum of five triangular PDFs. In other words, the sum of many triangular PDFs also produces a bell-shaped PDF. Following this argument further, the sum of any combination of many uniform PDFs and many triangular PDFs produces a bell-shaped PDF. This curiosity is getting intriguing. Let's try another example.

This time I'll deliberately start with PDF that's very "un-bell-like." The PDF shown in Figure 3.5 is about as un-bell-like as I can get. First of all, it looks like a V. I've even thrown a 0 probability point into the distribution. If anything, this is closer to an upside-down bell than it is to a bell. Second, it's asymmetric. How can adding up asymmetric PDFs lead to a symmetric PDF?

Figure 3.6 is the PDF for two of these "lopsided V distributions" added together. This still doesn't look very bell-like, but something suspicious is beginning to happen. Figure 3.7 is the PDF for four distributions added together. This is not exactly a nice smooth bell curve, but the features are

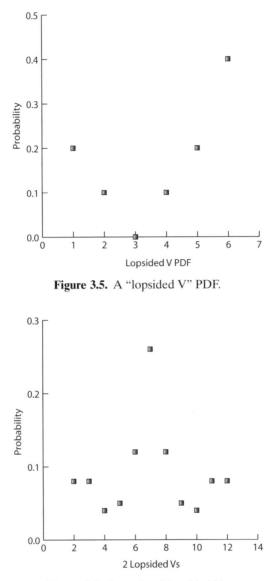

Figure 3.5. A "lopsided V" PDF.

Figure 3.6. Summing 2 lopsided Vs.

beginning to appear. Figure 3.8 is the PDF for 10 of these distributions added together. At this point there is no doubt what's happening—we're building another bell.

It takes many summations before the lopsided-V PDF turned into a bell-shaped curve. This is because I started out so very "un"-bell-shaped. Combinations of lopsided-V PDFs and uniform and/or triangular PDFs will look very bell-shaped much sooner.

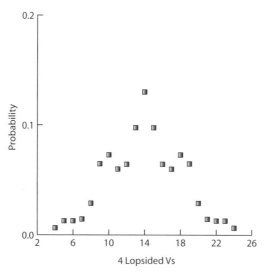

Figure 3.7. Summing 4 lopsided Vs.

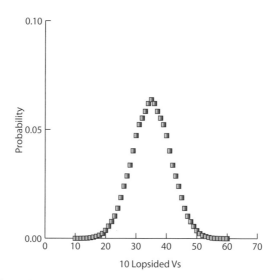

Figure 3.8. Summing 10 lopsided Vs: a bell-shaped curve.

By this time it should be clear that we're not looking at a coincidence. We're looking at one of the most intriguing mathematical relationships known. The formal name for this relationship is the Central Limit Theorem. Without getting into the precise mathematical language, this theorem says that the sums of the distributions of every PDF will, as we take more and more of these sums, approach a bell-shaped curve which is known as a normal, or Gaussian, distribution.

The Central Limit Theorem comes in two versions. The first is the "Fuzzy" Central Limit Theorem: When a collection of data comes about from many small and unrelated random effects, the data are approximately normally distributed. Think about it—normal distributions are everywhere: daily temperature swings, people's heights and weights, your average driving time to work when you start at about the same time every morning, students' grades on exams. This is the (in)famous Bell Curve that is so often quoted, invoked, praised, and damned.

The actual Central Limit Theorem is a bit more involved. Look back at the light bulb manufacturer—he's the guy who takes a bunch of light bulbs from his factory every day and runs them until they burn out. He averages the lifetime of this *sample* of his production and writes down the result in a little book. He repeats this process every day and then looks at the distribution of these averages, which we'll call the *sample mean*. The Central Limit Theorem says that if we have the size of these samples get larger and larger, this distribution approaches a Normal, or Gaussian, distribution. It also talks about the mean and standard deviation of the sample distribution and how they relate to the full population (of light bulb lifetimes), but we won't get into this.

Since the normal distribution is so important, we need to study it in some detail. Look at Figure 3.8 or Figure 3.4. The normal distribution is symmetric about its peak. That means that the mean equals the median equals the mode.

The full formula for the normal probability distribution function is

$$\text{normal distribution} = \frac{1}{\sqrt{2\pi}\sigma} e^{-(x-\mu)^2/2\sigma^2}$$

where μ is the mean and σ is the standard deviation.

This formula involves the exponent of e, which is the *base of natural logarithms*. This is a topic out of the calculus that isn't necessary for us to cope with, but I put the full formula in just for completeness.

The normal distribution as defined above is a continuous distribution. We can use it comfortably for discrete distribution problems; however, so long as there are enough x points for us to clearly recognize what we've got. Remember that our light bulb lifetimes are really a continuous distribution. We created a discrete distribution approximation by rounding lifetimes to the nearest hour.

The continuous normal distribution extends (in x) from minus infinity to plus infinity. There is therefore an issue when we say, for example, that your students' grades for a given semester "fit nicely onto a bell-shaped curve." How is it possible for there to be negative values of x allowed? How is it possible for there to be positive values of x so large that some student might have scored better than perfect on her exams?

This dilemma is resolved by kicking the tires of the term "approximate." The answer to this lies, as you shall see shortly, in the fact that once you're more than several standard deviations away from the mean in a normal distribution, the probability of an event is so low that we can (almost always) ignore it. The caveat here is that this answer doesn't always work. Sometimes the normal distribution is not correctly approximating reality and other distributions are needed. In Chapter 6 I'll present the Poisson distribution as an example of this. On the other hand, the (fuzzy) central limit theorem can be trusted—real situations arising from many small random distributions adding together look normal and if you just step back for a moment and look at what you're doing before proceeding you will never go wrong.

Figure 3.9 is the PDF of a normal distribution with a mean of 3 and a σ of 1. The vertical lines indicate the mean, 1σ and 2σ above and below the mean. Figure 3.10 is the CDF of the same normal distribution, with the same vertical lines. At 2σ below the mean, the value of the CDF is 0.0228. At 2σ above the mean, the value of the CDF is 0.9772, which is 1 − 0.9772 = 0.0228 down from 1.0. This means that the area outside these 2σ limits in the PDF are equal to 0.0228 + 0.0228 = 0.0456 of the total area of the PDF. Equivalently, the area inside these limits is equal to 1 − 0.0456 = 0.944. This is close enough to .095 = 95% that ±2σ about the mean is commonly accepted as the 95% confidence interval for the normal distribution. The 95% confidence interval is so often used as a reasonable confidence interval for everyday situations that it's often simply referred to as the "confidence interval."

The ±3σ range corresponds to approximately a 99.7% confidence interval, and a ±4σ range corresponds to a slightly more than 99.99% confidence interval. It's pretty clear that once we're more than 4σ away from the mean, this distribution is effectively 0 and it takes some pretty special circumstances to

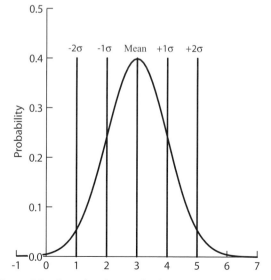

Figure 3.9. Gaussian (normal) distribution, μ = 3, σ = 1.

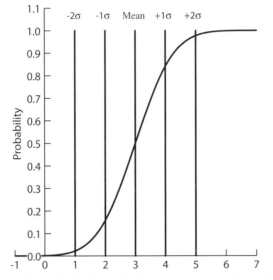

Figure 3.10. CDF of the distribution shown in Figure 3.9.

merit worrying about anything this far out. Consider both sides of this state-ment carefully. If we're looking at the distribution of weights of ears of corn or variations in annual rainfall or things like that, then we won't worry about anything outside the 99.99% confidence interval; we'll usually not worry about anything outside the 95% confidence interval. On the other hand, if we're looking at airline safety or drug safety issues, you can bet that we're examining *at least* the 99.99% confidence interval.

Oh, I forgot to mention that the ±1σ interval (about the mean) corresponds to about a 68% confidence interval. Also, remember that this is equivalent to saying that this ±1σ interval includes 68% of the points of a large sample of data that is normally distributed.

Before about 1985, before the explosion in the use of personal computers and inexpensive sophisticated pocket calculators, the only way the average person could try a calculation using confidence intervals of a normal distribu-tion was if she had a book of tables. Today this is not necessary—on a com-puter you could use a spreadsheet that has built-in normal distribution PDF and CDF calculators, or you could use a pocket calculator with statistical functions that also has these built-in calculators.

The CDF for a normal distribution is sometimes called the error function (ERF).

As an example, let's take a look at the average of a term's test scores of a class of 50 students. The tests were all graded on a 0 to 100% scale. Table 3.2 shows the list of scores, sorted in increasing order. I'd like to build a graph, called a *histogram*, of the distribution of these scores so we can *take a look* at

TABLE 3.2. Sample Test Scores

Test Scores	Test Scores	Test Scores	Test Scores
39.1	46.3	48.5	54.5
39.4	46.5	48.8	55.5
40.8	46.7	49.1	56.0
41.3	47.2	49.6	56.4
41.5	47.4	49.6	56.7
41.9	47.4	49.8	58.3
42.4	48.0	50.1	58.7
42.8	48.1	50.7	59.6
43.6	48.1	51.2	59.9
44.6	48.2	52.7	61.0
45.1	48.2	53.4	61.9
45.8	48.4	53.8	
46.1	48.4	54.3	

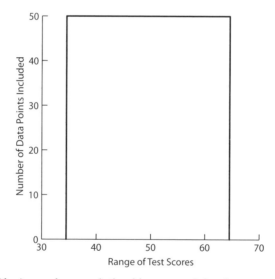

Figure 3.11. A very-low-resolution histogram of the class test scores data.

how they're distributed. (Remember in the last chapter I suggested that the best way to look at a PDF is to draw a graph and literally look at it.) What I have to do is to break this list into *bins*, each bin containing the test scores of a common chosen range. The question now is how to choose the range.

The test scores vary from a low of 39.1 to a high of 61.9. The average of the all the scores is 49.5. If I choose a bin that is centered at 49.5 and is 30 units wide, I get a graph that fits the definition of a histogram, but is really of no use whatsoever (Figure 3.11).

Now, the x axis shows test scores and the y axis shows the number of test scores included in each rectangle. Because the y axis isn't showing probabilities, this isn't literally a PDF. I'll get there shortly.

Going back to creating a histogram, I'll try bins 1 unit wide—for example, from 50.0 to 59.99. Figure 3.12 shows the resulting histogram. It's interesting, but I've gone too far the other way: Since there are only got 50 data points to work with, the bins are too narrow to be of much use.

I'll try bins 4 units wide centered at the average (49.5) (Figure 3.13). We could debate the aesthetics of this choice. We can, however, pretty clearly see the nature of the distribution now.

There is, as I just demonstrated, a bit of "art" in generating useful histograms. Also, unfortunately, this is one of the places where the "lies, damned lies, and statistics" curse can ring true. By deliberately making the bins too large or too small, it's often possible to hide something that you don't want to show or conversely to emphasize some point that fits your own agenda. You can't even claim that a person who does this is lying, because their truthful defense is that they did, indeed, use all of the data accurately when creating the histogram. The message here is that, when presented with a report where the conclusions will lead to or suppress some activity that you consider significant, get the data yourself and generate histograms using several different bin sizes so that you can draw your own conclusions about what the data is really telling you (or not telling you).

Next I would like to put some statistical numbers on our data set. Assuming we have a normal distribution, how do we get the best approximation to the

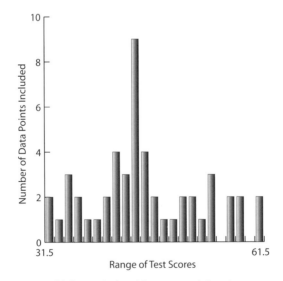

Figure 3.12. A too-high-resolution histogram of the class test scores data.

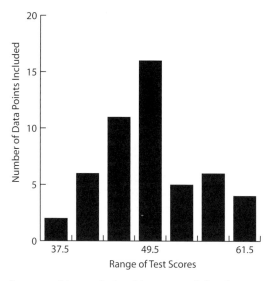

Figure 3.13. A reasonable resolution histogram of the class test scores data.

mean and the standard deviation? Remember that we won't get a really excellent fit because the data set is pretty small. There are two ways to do this. The first way is to use the formulas from Chapter 2:

$$\mu = \frac{1}{50}\sum_{i=1}^{50} x_i = \frac{1}{50}(39.1 + 39.4 + \cdots + 61.9) = 49.5$$

$$\sigma = \sqrt{\frac{1}{50}\sum_{i=1}^{50}(x_i - \mu)^2} = \sqrt{\frac{1}{50}[(39.1 - 49.5)^2 + \cdots + (61.9 - 49.5)^2]} = 5.8$$

At this point you might be a little suspicious—this isn't the formula for σ that I showed in Chapter 2. That formula had a weighting function. However, if you work your way backwards in Chapter 2, you'll see that the weighting function approach is useful when you have multiple instances of certain numbers or if you're dealing with a PDF weighting function. In this case we're dealing with the raw list of numbers so the formula looks a little different. Also, as you have probably already guessed, every list of numbers has a standard deviation—not just PDFs. And finally a last little point: The small correction described at the end of Chapter 2 only amounts to about a 1% error in the above formula, so I won't worry about it.

I made the claim above that $\pm 2\sigma$ about the mean in a normal distribution should encompass 95% of the data. In a relatively small sample, this is an approximation, but it should be a pretty good approximation. Now, $\mu - 2\sigma = 49.5 - 2(5.8) = 37.9$ and $\mu - 2\sigma = 49.5 + 2(5.8) = 61.1$. Looking at our list of numbers, this range encompasses everything but the highest number in the list (61.9). This is 49/50 = 98% of the data. Do we have a problem here? Actually, I generated this list of numbers from the normal distribution random number generator in a standard spreadsheet program using a mean of 50 and a standard deviation of 5. The problem is in the size of the sample.

Just how accurate an answer you can expect varies tremendously with sample size. This issue will be discussed further in the chapter on statistical sampling.

Returning to the class test scores, it's not uncommon for a teacher to use the statistics of test scores to determine grades. This is based on the belief that the average student should get the average grade and that test scores should fit pretty well onto a normal distribution (bell-shaped) curve. While these assertions might be reasonably true for a very large sample size, as we saw above they are only approximately true for a small group such as a class. In any case, a possible procedure (I don't think there are any hard and fast rules for doing this) would be to give students whose grades are within $\pm 1\sigma$ of the mean a C, grades between σ and 2σ above the mean a B, grades more than 2σ above the mean an A, grades between σ and 2σ below the mean a D, and grades more than 2σ below the mean an F. In a large enough sample size, this would imply that there would be 68% of the students getting a C, 13.5% getting a B, 2.5% getting an A, 13.5% getting a D, and 2.5% getting an F.

In many situations this procedure could give us utter nonsense. For example, suppose all of the grades are bunched between 49.0 and 51.0. We could still do the calculations and we could still allocate the grades, but the difference between highest-scoring student and the lowest-scoring student is so small that they should probably all get the same grade—although I'd be hard-pressed to say what that grade should be. In any case, arguing that a few of the students in this group should get an A while a few of them should fail simply doesn't make much sense. In a situation like this, you have to ask "Is the difference between a grade of 49.0 and a grade of 51.0, in a group of this size, statistically significant?" I'll discuss just what this means and how to calculate it in the chapter on statistical inference.

In upcoming chapters I will look at how situations that you wouldn't expect to produce bell-shaped curves (e.g., how far a drunken individual taking a "random walk" should be expected to travel or what to expect if you sit up all night tossing coins) bring us right back to the normal, or Gaussian, bell-shaped curve distribution.

CHAPTER 4

RANDOM WALKS

THE ONE-DIMENSIONAL RANDOM WALK

A random walk is a walk in which the direction and/or the size of the steps are random, with some PDF(s). In two or three dimensions the random walk is sometimes called Brownian motion, named after the botanist Robert Brown, who described seemingly random motion of small particles in water. In this chapter I'll tie together the idea of a random walk—which is another example of how the sum of experiments with a very simple PDF start looking normal—with the physical concept of *diffusion*, and I'll take a deeper look into what probability really means.

Let's imagine that you're standing on a big ruler that extends for miles in both directions. You are at position zero. In front of you are markings for +1 foot, +2 feet, +3 feet, and so on. Behind you are markings for −1 foot, −2 feet, and so on. Every second you flip a coin. If the coin lands on a head you take a step forward (in the + direction). If the coin lands on a tail, you take a step backward (in the − direction). We are interested in where you could expect to be after many such coin flips.

If you were to keep a list of your coin flips, the list would look something like:

Probably Not: Future Prediction Using Probability and Statistical Inference,
by Lawrence N. Dworsky.
Copyright © 2008 John Wiley & Sons, Inc.

1. Head. Take 1 step forward.
2. Tail. Take 1 step backward.
3. Head. Take 1 step forward.
4. Head. Take 1 step forward.
5. Head. Take 1 step forward.
6. Tail. Take 1 step backward.
7. And so on.

Now, assign a value of +1 to each step forward (heads) and a value of −1 to each step backward (tails), then your position after some number of coin flips would be just the algebraic sum of each of these values. For example, from the list above, you would have (+1 −1 +1 +1 +1 −1) = +2. Figure 4.1 shows the results of a typical walk of this type. Random walks can in general have more complex rules. The size of each step can vary according to some PDF, the time between steps can vary, and of course a real walk would have to be described in two dimensions, with the direction of each step also being random. This simple random walk, however, is complex enough for us right now.

There are three comments I want to make here:

1. Since heads and tails are equally likely, you would expect the sum to be about 0. In other words, you wouldn't really expect to get anywhere.
2. If you were gambling with the rules that you win $1 for each head and lose $1 for each tail rather than stepping forward and backward, this same procedure would tell you exactly your winnings (or losses) after a number of coin flips. This is, in other words, analogous to the simple coin flip gambling game.

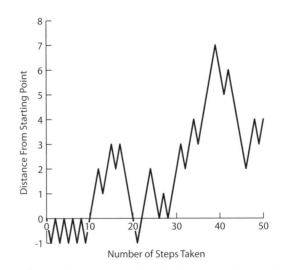

Figure 4.1. Example of a simple 50-step random walk.

3. Since we're looking at the sum of the results of a number of statistical events, we should suspect that something normal will be raising its head soon.

After the first coin flip, you would have moved either 1 step forward or 1 step backward, with an equal (50%) probability of either. The PDF of this situation, Figure 4.2, is quite simple. It impossible at this time for you to be at your starting point. For that matter, after any odd number of coin flips it is impossible for you to be at the starting point, again directly analogous to the coin flip gambling game.

Continuing, Figure 4.3 shows the PDF for your position after 2 coin flips; it is just the sum of two of PDFs of Figure 4.2. Figures 4.4, 4.5, and 4.6 are the PDFs after 3, 4, and 5 steps, respectively.

It seems that we are once again *building a bell*. In this example the mean value, μ, is identically 0 regardless of the number of coin flips.

Now let's start looking at probabilities. First consider the probability of ending up just where you started (at $x = 0$). Since this probability is always identically 0 for an odd number of coin flips, I'll just present the even cases. After n coin flips, the probability of being at $x = 0$ is shown in Table 4.1. The more times you flip the coin, the less likely it is that you'll end up where you started (at $x = 0$).

What about landing near $x = 0$, say at $x = 2$ or $x = -2$? Table 4.2 is a repeat of Table 4.1 with this new probability added to it.

While the probability of being at $x = 0$ starts at 1 and always decreases (with increasing n), the probability of being at $x = 2$ starts at 0, increases to a peak,

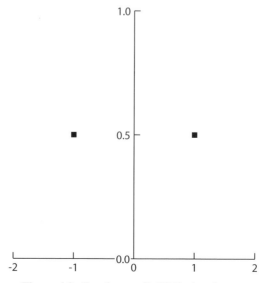

Figure 4.2. Random walk PDF after 1 step.

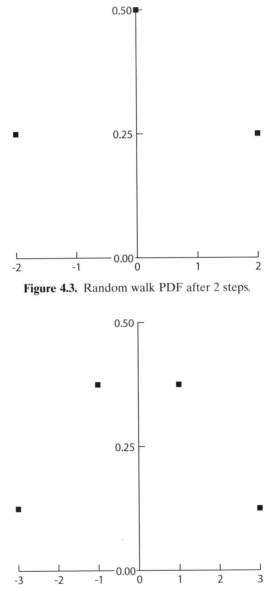

Figure 4.3. Random walk PDF after 2 steps.

Figure 4.4. Random walk PDF after 3 steps.

and then starts decreasing. This makes sense—before you flip the coin, it is impossible to be anywhere but at $x = 0$; after many coin flips there is a large opportunity to be in different places.

Comparing the probabilities of being at $x = 0$ and at $x = 2$, let's look both at the ratio of these probabilities and also their absolute difference. Table 4.3 is an extension of Table 4.2, with these two new columns added.

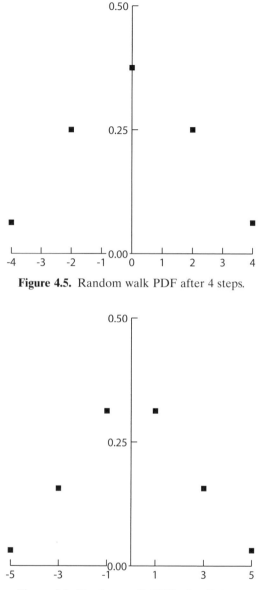

Figure 4.5. Random walk PDF after 4 steps.

Figure 4.6. Random walk PDF after 5 steps.

Both of these latest calculations are telling us the same thing. The ratio of the two numbers is approaching 1 and the difference between the two numbers is approaching 0 as n gets larger. In other words, for very large values of n, we can expect the probability of finding yourself at $x = 0$ and at $x = 2$ to be almost the same—even though these probabilities are themselves both falling as n grows. This means that even though the expected distance from $x = 0$ is

TABLE 4.1. Probability of Landing at $x = 0$ Versus Number of (Random) Steps

N	Probability of $x = 0$
0	1
2	.375
4	.3125
6	.273
8	.246
10	.226
. . .	. . .
30	.144

TABLE 4.2. Same as Table 4.1, with $x = 2$ Added

N	Probability of $x = 0$	Probability of $x = 2$
0	1	0
2	.375	.250
4	.3125	.250
6	.273	.235
8	.246	.219
10	.226	.215
. . .	. . .	. . .
30	.144	.135

TABLE 4.3. Same as Table 4.2, with Ratio & Difference Columns Added

N	Probability of $x = 0$	Probability of $x = 2$	Ratio	Difference
0	1	0	0	1
2	.375	.250	.667	.125
4	.3125	.250	.800	.063
6	.273	.235	.861	.038
8	.246	.219	.890	.027
10	.226	.215	.951	.011
. . .	. . .	. . .	. . .	. . .
30	.144	.135	.968	.009

always 0, the more steps you take the less likely it is that you'll be at $x = 0$ or anywhere near it.

As a last calculation, let's look at the standard deviation of the values of x. This calculation is a direct repeat of the examples in earlier chapters, so I won't show the details here. I'll eliminate the two right-hand columns in Table 4.3, just for clarity, and show σ in Table 4.4. Figure 4.7 shows how σ varies with n.

TABLE 4.4. How σ Varies with N

N	σ
0	
2	1.414
4	2.000
6	2.449
8	2.828
10	3.162
...	...
30	5.477

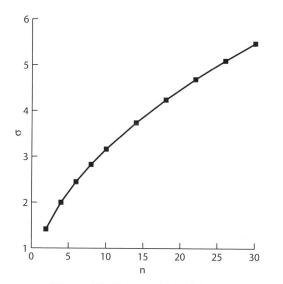

Figure 4.7. How σ varies with n.

σ, it seems, gets larger as n gets larger, although its rate of growth with respect to n slows as n grows. If you look at the numbers carefully, you'll quickly see that the functional relationship is simply $\sigma = \sqrt{n}$.

Let's summarize the situation: While we know that a large number of coin flips should yield, in some sense, an approximately equal number of heads and tails, we seem to be at a loss as to how to put our fingers on in just what sense this is happening. By every measure so far, the more times we flip the coin, the less likely it is that we'll see an equal number, or even an approximately equal number, of heads and tails.

All, at last, is not lost. After n coin flips, it is possible to have gotten n heads (or n tails). In terms of the random walk, this is equivalent to saying that it's possible to be n steps away from the starting point, in either direction. The standard deviation of the PDF, however, only increases with the square root

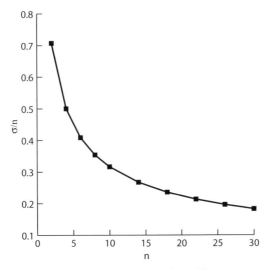

Figure 4.8. How σ/n varies with n.

of n. Therefore, if we look at the ratio of sigma to n, we get Figure 4.8. Doing a little bit of arithmetic,

$$\frac{\sigma}{n} = \frac{\sqrt{n}}{n} = \frac{1}{\sqrt{n}}$$

As Figure 4.8 shows, the ratio σ/n is getting smaller as n gets larger. From the last chapter, recall that for a normal distribution, the region inside ±2σ about the mean encompasses 95% of the event probabilities; this is called the "95% confidence interval." n is the farthest distance you could have walked (or the highest number of heads or tails that you might have flipped). Therefore, if σ/n is getting smaller than the region in which you are 95% confident of ending up after n steps, it is getting to be a smaller *fraction* of the furthest distance that you might possibly have ended up.

A couple of ancillary points:

1. This same argument holds for any confidence interval (number of σ). In other words, the region about your starting point in which you are any % confident of ending up after n steps is getting to be a smaller fraction of how far away you might possibly be.

2. Don't forget that σ is increasing as n increases. The 95% (or another %) wide confidence factor region is getting larger as n increases. For a large value of n, you really have no idea where you'll be—you just know that it's highly unlikely that you'll be as far away as you possibly could be.

The above points pass common-sense tests: If you flip a coin 4 times and get 3 heads, it's certainly not a big deal. If you flip a coin 400 times and get 300 heads, however, this doesn't look like an honest coin and the random walk based on these flips starts looking very directed. Similarly, if you flip a coin 20 times and get 18 more heads than tails (i.e., 19 heads and 1 tails), you should be suspicious. If you flip a coin 200 times and get 18 more heads than tails (109 heads and 91 tails), however, it's really no big deal. The exact probabilities of the above events can be calculated by using the binomial probability formula, which I'll be talking about in another chapter, or by using a graphical technique that I'll show below.

WHAT PROBABILITY REALLY MEANS

From Figures 4.7 and 4.8 we now have a good picture of what *probability* really means: We have already seen that when we flip a coin n times (n even), the highest probability, or most likely event, is that we'll get $n/2$ heads and $n/2$ tails. As we increase n (i.e., flip the coin more and more times), even though it's less likely in an absolute sense that we'll get $n/2$ heads and $n/2$ tails, the ratio of the standard deviation to n falls. In other words, the confidence interval, for any measure of confidence, divided by n, falls. Putting this in terms that are (hopefully) easier to digest, the confidence interval as a percentage (fraction) of the number of coin flips falls as n increases.

Relating this to the random walk, the confidence interval about the expected location is telling us where (in what range of x) on the ruler we have a 95% probability of being. As n gets larger (more steps), both the standard deviation of x and the confidence interval grow. However, the confidence interval *as a fraction of n* falls.

The point I'd like to make here can be skipped over if you're uncomfortable with the algebra: Using just a little bit of algebra, I can derive some very powerful conclusions about what we can expect based upon the number of coin flips. If I were to require that the 95% confidence interval be no more than 10% of the number of coin flips, then I would require that

$$2\sigma = 2\sqrt{n} \le .1n$$

Carrying through the algebra, I find that $n \ge 400$. If I wanted the 95% confidence interval to be no more than 1% of the number of coin flips, then using the same calculation, I get $n \ge 40,000$.

Summarizing, if you want to know something to a given confidence interval as a very small fraction of the number of data points you've got, you need a very large number of data points.

From the perspective of the graph of a normal PDF, think of it this way: If I were to fix the size on a piece of paper of steps in n so that, for example, a graph showing the $n = 4$ steps (flips) case took 1 inch, then a graph for the

$n = 16$ steps (flips) would take 4 inches and you would see that σ is bigger when n is bigger. In other words, the distribution PDF is much fatter for the $n = 16$ case than for the $n = 4$ case. On the other hand, if I drew both graphs so that they were 4 inches wide (i.e., scaled the x axis by n), the PDF for the $n = 16$ case would look much narrower than the graph for the $n = 4$ case. Both of these perspectives are shown in Figure 4.9 (using lines rather than points because the graph is very busy).

Table 4.5 shows an interesting way to look at (or calculate) the probabilities of this random walk. At the start, our walker is at $x = 0$. He hasn't gone anywhere yet, so his probability of being at $x = 0$ is 1.000. He takes his first step, either to the left or the right, with equal probabilities. This means that the probability of his being at $x = 1$ or $x = -1$ must be half of his previous probability, while his probability of being where he was ($x = 0$) is exactly 0. The is shown in the row labeled *step 1*.

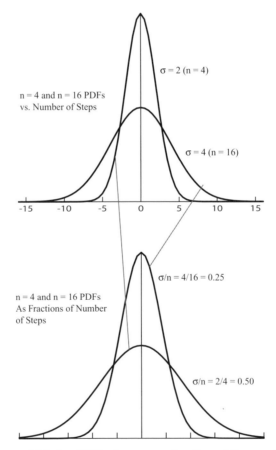

Figure 4.9. Four- and 16-step PDFs shown as a function of the number of steps and as a fraction of the number of steps.

TABLE 4.5. Triangle of "Propagating" Probabilities

Step					Probability				
0					1.000				
1				0.500	0.	0.500			
2			0.250	0.	0.500	0.	0.250		
3		0.125	0.	0.375	0.	0.375	0.	0.125	
4	0.0625	0.	0.250	0.	0.375	0.	0.250	0.	0.0625
Position:	−4	−3	−2	−1	0	1	2	3	4

Getting to step 2 is slightly, but not very much, more complicated. From each of the positions he might have been at after step 1, he has half of his previous probability of moving 1 step to the right or 1 step to the left. Therefore his probability of being at $x = 2$ or $x = -2$ is half of 0.500, or 0.250. Since he has 2 chances of arriving back at position 0, each with probability 0.250, his probability of getting to position 0 is 0.250 + 0.250 = 0.500. Again, his probabilities of being where he had been at step 1 are exactly 0.

This can be carried on forever. In each case, the probability at a given spot "splits in half" and shows up down one line and both one step to the left and one step to the right. When a point moves down and, say, to the right and lands in the same place as another point moving down and to the left, then the two probabilities add.

DIFFUSION

The above process can be continued ad infinitum (which I think is Latin for *until we get bored*). When viewed this way, we may think of the probability as some sort of propagating phenomenon, *smearing out* across space with time. This perspective ties the idea of a random walk to a physical process called *diffusion*. I'll illustrate this now with a slightly different perspective on the random walk process.

At time = 0 I'll deposit a very large number of ants in a starting region. I'll put 10 million ants at $x = 0$, 10 million ants at $x = -1$, and 10 million ants at $x = +1$. I'll assume that each ant wanders either 1 unit in the $+x$ or 1 unit in the $-x$ direction per second randomly, with an equal probability of going in either direction.

My ant distribution function (ADF) at time = 0 is shown in Figure 4.10, and it is shown at time = 1 in Figure 4.11.

At time = 0, there were 10 million ants at $x = -1$, $x = 0$, and $x = 1$ because I put them there. Between time = 0 and time = 1, there was an average motion of 5 million ants in both the $+x$ and $-x$ directions from each of these three locations. This means that there was no net flux of ants into or out of $x = 0$. The only locations that see a net flux in or out are those locations where there

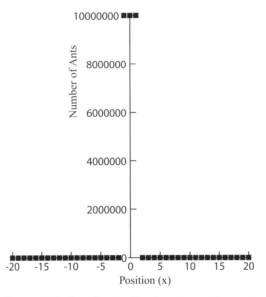

Figure 4.10. Ant distribution function at time = 0.

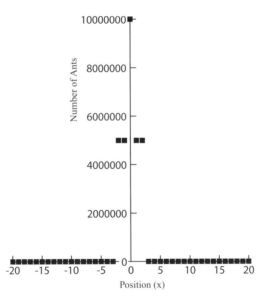

Figure 4.11. ADF at time = 1.

is a difference in the starting numbers between those locations and adjacent locations. This process is known as *diffusion*.

There is only a net flow in any direction at a given point due to diffusion when there is a difference in the concentration of particles (i.e., a *concentration gradient*) on either side of the point. The net flow must always be in the direction from the higher concentration to the lower concentration. Diffusion occurs in gases and in liquids (and at very high temperatures in solids). Particles in a liquid and gas molecules have random velocities (naturally with some distribution function) due to thermal agitation. When we can see small particles randomly moving around in a host liquid, we call the process *Brownian motion*.

Continuing with our example, Figure 4.12 shows what we have at time = 19.

The particles are spreading out, moving (or diffusing) away from their starting cluster and trading their original uniform density for the familiar normal distribution. This is what happens when a few drops of cream are dripped into a cup of coffee, when some colored gas is introduced into a colorless gas, and so on.

In Figure 4.12 you can see an artifact of the "roughness" of our numerical approximation. In real life, the ants wouldn't conveniently group themselves into uniform bunches at $x = 0$, $x = 1$, and so on. They would be in bunches about 1 ant wide, each moving randomly, and the curve would smooth out to an ideal normal distribution curve. In the graphs below, I'll do this smoothing just to make things easier to look at.

Next, I'll modify the situation by putting walls at $x = 20$ and $x = -20$. These walls have the property of turning around, or bouncing off, any ant (particle) that crashes into them. For time < 20 this has no effect whatsoever, because no ants have reached the walls yet. This is the situation at time = 19.

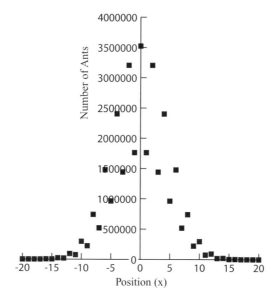

Figure 4.12. ADF at time = 19.

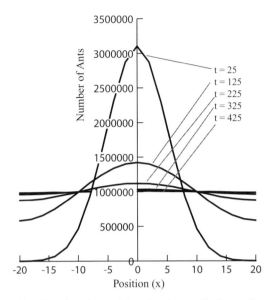

Figure 4.13. ADF in a box with reflecting walls for various times.

Figure 4.13 shows the situation at $t = 25$, which is hardly different from the situation at $t = 19$, and then at several later times, in time steps of 100. The discrete points of the above graphs have been replaced with continuous lines for clarity.

The distribution of ants in the box, or gas molecules in a box, or cream in a cup of coffee, or . . . , diffuses from a high uniform concentration over a small region ($t = 0$) to a lower uniform concentration over the entire box. The total number of ants, or . . . , of course remains constant. In Chapter 16 it will be shown how the concepts of gas temperature, gas pressure, and the ideal gas laws derive from this simple calculation (extended of course into three dimensions). This is a characteristic we see over and over again; cooking smells permeate the house.

As a last point, let me add some information that really belongs in Chapter 2 as a property of standard deviations, but is easier to explain with Chapters 3 and 4 behind you rather than still in front of you.

In a random walk, the steps are independent of each other. That is, each step is determined by (the equivalent of) the flip of a coin, in which case the next step has no knowledge whatsoever of what the previous steps were. The standard deviation, σ in this situation, grows with the square root of the number of steps, $\sqrt{n}$.

This square root relationship is a general property of independent (random) events. For example, suppose I have an old clock with somewhat warped parts. The clock, on the average, keeps pretty good time—but at any given time, its hands will predict a time that's distributed normally about the actual time, with a standard deviation of 3 minutes. My sister has a similar clock, but with even worse warpage. Her clock will also predict a time that's distributed normally about the actual time, but in her case the standard deviation is 4 minutes. If I compare the clocks every 5 minutes for a full day, what will the distribution of the differences of the predicted times of these clocks be?

Since both clocks, on the average, tell the correct time, the average value of the difference between them will be 0. The standard deviation of these errors, however, will be

$$\sigma_{\text{difference}} = \sqrt{3^2 + 4^2} = \sqrt{9 + 16} = \sqrt{25} = 5$$

This is a smaller answer than you get by just adding $3 + 4 = 7$. This is because both clocks' errors are random, and there is some propensity to cancel. It is directly analogous to the random walk in that since forward and backward steps are equally likely, they'll tend to cancel (that's why the expected value of distance traveled is 0) but the standard deviation keeps growing with the number of steps, and it grows as the square root of the number of steps.

If the steps, or clocks, or whatever, are not independent, then all bets are off. For example, if by some measure our random walker is more likely to take a step in the same direction as his previous step because he's walking on a hill, then the steps are not independent. In this case it's impossible to predict the standard deviation without knowing the exact details of the dependence (i.e., the non-independence) of a step on the previous step (or steps).

CHAPTER 5

LIFE INSURANCE AND
SOCIAL SECURITY

Having just burrowed through a couple of chapters of probability concepts and definitions without any real examples of how this *stuff* affects real-world situations (that is, unless you consider watching ants take random walks a real-world situation), it's time to start putting this knowledge to use. Insurance in its various forms is something that just about all of us deal with at some time or other. There are so many different kinds of insurance policies that it's impossible for me to even consider discussing them all. I chose to concentrate on simple life insurance policies. Most people don't actually buy simple life insurance policies—insurance companies like to put together combinations of life insurance policies, savings accounts, annuities, and so on, into big packages. There's nothing inherently wrong with them doing this—provided you understand what it is you're buying, what the terms of each of the components of the package are, and how all of these components interact.

There are also different forms of simple life insurance policies as well as different names for the same life insurance policy. The simplest life insurance policy, which will be discussed in detail below, is the *term* life insurance policy. When you buy a simple term policy, you pay a certain amount up front (the *premium*). If you die during the specified term of the policy, the insurance company pays a fixed amount that is specified in the policy. If you survive the term of the policy, you and the insurance company no longer have a relationship—they keep the premium, and they owe you nothing ever unless of course you purchase another policy. A simple variation of the term policy is the *decreasing* term policy. You pay a premium up front and if you die during the term of the policy, the insurance company pays off. The amount of the payoff, however, decreases in steps over the term of the policy. A renamed decreasing term policy is a mortgage insurance policy. The payoff amount of the mortgage insurance policy is tied to the balance of the mortgage on your home which decreases as you make your monthly mortgage payments. Also, you typically pay the premium periodically rather than up front. I could keep going here, but I think these definitions give you enough information to start looking at the probability calculations involved.

Real insurance companies, like all other real companies, are in business to make money. Along the way, they have a cost of doing business which comes not just from paying out on policies but also from salaries, taxes, office equipment, real estate, and so on. The simple rule is that they must take in more than they pay out to policy holders in order for them to stay in business. Business finance is a topic well worth studying (especially if you're running or planning to run a business). This book, however, is about probability and this chapter is about how probability calculations determine life insurance policy costs. I'll therefore leave the business issues to books about business issues. My insurance companies are idealized operations staffed by volunteers. They have no costs other than policy payouts, they make no profits, and they take no losses. The expected value of their annual cash flow is zero.

INSURANCE AS GAMBLING

Suppose that you walk into your bookie's "office" and ask: "What are the odds that I will die of a cause other than suicide in the next 12 months?" The bookie knows you to be fairly young and healthy, and he's sure that you have no diseases that you're hiding. He stares at the ceiling for a while and/or consults his crystal ball and replies: "99 to 1 odds against it."[1] Let's assume that the bookie is willing to take this bet: You give him $1000. He gives you his *marker*, which is a note that says something like "If you die for any reason (other than suicide) within 12 months of today, I will give $100,000 to you—actually to your family, considering the nature of the bet. In either case, I keep the $1000."

Most of us really do handle things this way, we just change the names of the items and the title of our bookie. The more common scenario is for you to walk into your insurance broker's office and ask: "What are the odds that I will die in the next 12 months?" Your insurance broker asks you to fill out some forms, possibly to get a physical examination. He then stares at his Actuarial, or Life, Tables for a while and replies "99 to 1 odds against it."

---

[1] Just in case you've forgotten the definition: 99 to 1 odds against you dying means that the probability of you dying is $1/(99 + 1) = 1/100 = 0.01$.

Once again, we'll assume that he is willing to take the bet—that is, to write you a policy. You give him $1000. He gives you *his marker*, which in this case is labeled "Term Life Insurance Policy." The insurance policy, boiled down to its simplest terms, says something like "If you die for any reason other than suicide within 12 months of today, I will give $100,000 to you—actually, to people called "beneficiaries," considering the nature of the policy. Again, whether you die during this period or not, the insurance company keeps the $1000.

When looked at this way, it becomes clear that both gambling casinos and insurance companies are in the same business—taking long odds bets from people who come through the door. Gambling casinos usually deal with slot machine kinds of bets while insurance companies usually deal with insurance kinds of bets, but they both have very sophisticated sets of tables telling them the odds on every bet that they might take. In the case of the gambling casino, these tables take the form of probabilities based on the mechanisms in the slot machines and the numbers layout of the roulette wheels; in the case of the insurance company, these tables take the form of average life expectancies; and so on. Since terms such as *betting* and *gambling* tend to make people nervous, whereas *having insurance* tends to make people feel warm and secure, insurance companies have their own language and are careful to minimize any references to betting and odds. From the point of view of the casino and the insurance company the difference between the two different types of bets is, of course, is totally cosmetic.

Let's examine the purchase of a life insurance policy. Assuming that you're in reasonable mental and physical health and are living a fairly happy life, you really don't want to win this bet. That is, you don't want to die sometime in the next 12 months. Fortunately, in the above example, the odds are 99 to 1 against you dying in the next 12 months.

We must now confront a very perplexing situation: You have placed a bet where the odds are 99 to 1 against you winning; and furthermore, **you really don't want to win this bet**. (Remember that to win this bet you must die sometime in the next 12 months.) From this perspective, placing the bet doesn't seem like a good idea at all. Why then did you place this bet? Why do so many of us place this bet—that is, buy life insurance policies?

The answer lies in a personal application of a philosophy called the *Precautionary Principle*: A reasonable person will bet against very long odds, on a bet that he really does not want to win, if the side effects of winning would be absolutely unacceptable. In the case of life insurance, absolutely unacceptable could mean that if the wage earner of a family dies, the family would not have the financial resources to carry on. Having a life insurance policy to protect against this situation is a very good idea.

The Precautionary Principle as you would find it in many sites on the internet usually deals with scientific and/or political decisions affecting public issues such as the environment and public safety. There are many definitions cited and even discussions comparing and contrasting these different definitions.

An example of these definitions is "... a willingness to take action in advance of scientific proof [or] evidence of the need for the proposed action on the grounds that further delay will prove ultimately most costly to society and nature, and, in the longer term, selfish and unfair to future generations."[2] It's not hard to imagine how this societal definition evolved from a personal definition such as the one I gave above and how the personal definition in turn evolved from the pattern-generating "there might be tigers in the bush" response discussed in Chapter 1.

Getting back to the difference between gambling and buying insurance, our first conclusion is that from the perspective of the gambling casino owner and the insurance company, there is absolutely no difference. In both cases there are a lot of customers so it's unlikely that there will be too much deviation from the actuarial tables/odds. Each costumer is paying a small amount as compared to the payback—another way of saying that these are "long odds." In practice of course the gambling casino and the insurance company skews the price:payoff ratio somewhat so as to assure a steady flow of money into their pockets to cover costs of running the system and, of course, for profits. In both cases there are lots of attractive products on sale, so it's pretty certain that each customer will in reality pay more than a "small amount." In the case of the gambling casino there are slot machines, roulette wheels, craps tables, and so on. In the case of the insurance company there is health insurance, auto insurance, and so on, and of course the many names for the life-insurance-based products.

Should we conclude that there's absolutely no difference between placing a gambling casino bet and buying an insurance policy? No. There is one very big difference which is implicit in the discussion above. When you put money into a slot machine, you're hoping to win the bet. When you pay a year's premium on a life insurance policy, you're hoping to lose the bet but cannot accept the consequences of winning the bet unless you have placed a big enough bet. In other words, you're hoping you don't die during the next 12 months, but just in case you do die, you want to make sure you have enough life insurance that your heirs are left economically secure. This simple but very important difference makes gambling a frivolous and sometimes painful entertainment, while prudent insurance purchases are an important part of personal and family financial security and health planning.

LIFE TABLES

In order to be able to calculate the average costs of writing life insurance policies, we need to know the probabilities of people dying at different ages. This information is presented in Life Tables. Life Tables are available on the

---

[2] Interpreting the Precautionary Principle, edited by Tim O'Riordan and James Cameron, Earthscan Publications Ltd., 1994.

web, along with tutorials on how they're calculated, precise definitions of the terms used, and so on. Tables are calculated and published for various subgroups of the population such as sex and race as well as for the entire population. These tables must be updated frequently because advances in medical care, nutrition, demographics, and so on, cause changes in the data. Table 5.1 shows the data for the entire population of the United States in the year 2003. Note that the third column from the left, "Number Surviving to Age x", starts with 100,000 people. Obviously this number has nothing to do with the population of the United States in 2003. It is just a convenient starting number from which we can calculate fractions of the population; that is, if we have 99,313 people this year and had 100,000 people last year, then 99313/100000 = 99.313% of the people who were around last year are still around this year.

The first column (on the left) is age, in one-year intervals. At the bottom of the table you can see that there are no specific data for people more than

TABLE 5.1. 2003 Life Table for the Entire U.S. Population

Age	Probability of Dying Between Ages x and $x + 1$, q_x	Number Surviving to Age x, l_x	Number Dying Between Ages x and $x + 1$, d_x	Person-Years Live Between Ages x and $x + 1$, L_x	Total Number of Person-Years Lived Above Age x, T_x	Expectation of Life at Age x, e_x
0–1	0.006865	100,000	687	99,394	7,748,865	77.5
1–2	0.000465	99,313	46	99,290	7,649,471	77.0
2–3	0.000331	99,267	33	99,251	7,550,181	76.1
3–4	0.000259	99,234	26	99,222	7,450,930	75.1
4–5	0.000198	99,209	20	99,199	7,351,709	74.1
5–6	0.000168	99,189	17	99,181	7,252,510	73.1
6–7	0.000151	99,172	15	99,165	7,153,329	72.1
7–8	0.000142	99,158	14	99,150	7,054,164	71.1
8–9	0.000139	99,143	14	99,137	6,955,013	70.2
9–10	0.000134	99,130	13	99,123	6,855,877	69.2
10–11	0.000165	99,116	16	99,108	6,756,754	68.2
11–12	0.000147	99,100	15	99,093	6,657,646	67.2
12–13	0.000176	99,085	17	99,077	6,558,553	66.2
13–14	0.000211	99,068	21	99,057	6,459,476	65.2
14–15	0.000257	99,047	25	99,034	6,360,419	64.2
15–16	0.000339	99,022	34	99,005	6,261,385	63.2
16–17	0.000534	98,988	53	98,962	6,162,380	62.3
17–18	0.000660	98,935	65	98,903	6,063,418	61.3
18–19	0.000863	98,870	85	98,827	5,964,516	60.3
19–20	0.000925	98,784	91	98,739	5,865,689	59.4
20–21	0.000956	98,693	94	98,646	5,766,950	58.4
21–22	0.000965	98,599	95	98,551	5,668,304	57.5
22–23	0.000987	98,504	97	98,455	5,569,753	56.5

TABLE 5.1. *Continued*

Age	Probability of Dying Between Ages x and x + 1, q_x	Number Surviving to Age x, l_x	Number Dying Between Ages x and x + 1, d_x	Person-Years Live Between Ages x and x + 1, L_x	Total Number of Person-Years Lived Above Age x, T_x	Expectation of Life at Age x, e_x
23–24	0.000953	98,406	94	98,360	5,471,298	55.6
24–25	0.000955	98,313	94	98,266	5,372,938	54.7
25–26	0.000920	98,219	90	98,174	5,274,672	53.7
26–27	0.000962	98,128	94	98,081	5,176,499	52.8
27–28	0.000949	98,034	93	97,987	5,078,418	51.8
28–29	0.000932	97,941	91	97,895	4,980,430	50.9
29–30	0.000998	97,850	98	97,801	4,882,535	49.9
30–31	0.001014	97,752	99	97,703	4,784,734	48.9
31–32	0.001046	97,653	102	97,602	4,687,032	48.0
32–33	0.001110	97,551	108	97,497	4,589,430	47.0
33–34	0.001156	97,443	113	97,386	4,491,933	46.1
34–35	0.001227	97,330	119	97,270	4,394,547	45.2
35–36	0.001357	97,210	132	97,145	4,297,277	44.2
36–37	0.001460	97,079	142	97,008	4,200,132	43.3
37–38	0.001575	96,937	153	96,861	4,103,124	42.3
38–39	0.001672	96,784	162	96,703	4,006,264	41.4
39–40	0.001847	96,622	178	96,533	3,909,561	40.5
40–41	0.002026	96,444	195	96,346	3,813,027	39.5
41–42	0.002215	96,249	213	96,142	3,716,681	38.6
42–43	0.002412	96,035	232	95,920	3,620,539	37.7
43–44	0.002550	95,804	244	95,682	3,524,620	36.8
44–45	0.002847	95,559	272	95,423	3,428,938	35.9
45–46	0.003011	95,287	287	95,144	3,333,515	35.0
46–47	0.003371	95,000	320	94,840	3,238,371	34.1
47–48	0.003591	94,680	340	94,510	3,143,531	33.2
48–49	0.003839	94,340	362	94,159	3,049,021	32.3
49–50	0.004178	93,978	393	93,782	2,954,862	31.4
50–51	0.004494	93,585	421	93,375	2,861,080	30.6
51–52	0.004804	93,165	448	92,941	2,767,705	29.7
52–53	0.005200	92,717	482	92,476	2,674,764	28.8
53–54	0.005365	92,235	495	91,988	2,582,288	28.0
54–55	0.006056	91,740	556	91,462	2,490,300	27.1
55–56	0.006333	91,185	577	90,896	2,398,838	26.3
56–57	0.007234	90,607	655	90,279	2,307,942	25.5
57–58	0.007101	89,952	639	89,632	2,217,662	24.7
58–59	0.008339	89,313	745	88,941	2,128,030	23.8
59–60	0.009126	88,568	808	88,164	2,039,089	23.0
60–61	0.010214	87,760	896	87,312	1,950,925	22.2
61–62	0.010495	86,864	912	86,408	1,863,614	21.5
62–63	0.011966	85,952	1029	85,438	1,777,206	20.7

TABLE 5.1. *Continued*

Age	Probability of Dying Between Ages x and x + 1, q_x	Number Surviving to Age x, l_x	Number Dying Between Ages x and x + 1, d_x	Person-Years Live Between Ages x and x + 1, L_x	Total Number of Person-Years Lived Above Age x, T_x	Expectation of Life at Age x, e_x
63–64	0.012704	84,923	1079	84,384	1,691,768	19.9
64–65	0.014032	83,845	1177	83,256	1,607,384	19.2
65–66	0.015005	82,668	1240	82,048	1,524,128	18.4
66–67	0.016240	81,428	1322	80,766	1,442,080	17.7
67–68	0.017837	80,105	1429	79,391	1,361,314	17.0
68–69	0.019265	78,676	1516	77,918	1,281,923	16.3
69–70	0.021071	77,161	1626	76,348	1,204,004	15.6
70–71	0.023226	75,535	1754	74,658	1,127,657	14.9
71–72	0.024702	73,780	1823	72,869	1,052,999	14.3
72–73	0.027419	71,958	1973	70,971	980,130	13.6
73–74	0.029698	69,985	2078	68,946	909,159	13.0
74–75	0.032349	67,906	2197	66,808	840,213	12.4
75–76	0.035767	65,710	2350	64,535	773,405	11.8
76–77	0.039145	63,360	2480	62,119	708,870	11.2
77–78	0.042748	60,879	2602	59,578	646,751	10.6
78–79	0.046289	58,277	2698	56,928	587,172	10.1
79–80	0.051067	55,579	2838	54,160	530,244	9.5
80–81	0.056846	52,741	2998	51,242	476,084	9.0
81–82	0.061856	49,743	3077	48,204	424,842	8.5
82–83	0.067173	46,666	3135	45,099	376,638	8.1
83–84	0.077268	43,531	3364	41,850	331,539	7.6
84–85	0.079159	40,168	3180	38,578	289,689	7.2
85–86	0.086601	36,988	3203	35,386	251,112	6.8
86–87	0.094663	33,785	3198	32,186	215,725	6.4
87–88	0.103381	30,587	3162	29,006	183,539	6.0
88–89	0.112791	27,425	3093	25,878	154,534	5.6
89–90	0.122926	24,331	2991	22,836	128,656	5.3
90–91	0.133819	21,340	2856	19,913	105,820	5.0
91–92	0.145499	18,485	2689	17,140	85,907	4.6
92–93	0.157990	15,795	2495	14,547	68,767	4.4
93–94	0.171312	13,300	2278	12,160	54,220	4.1
94–95	0.185481	11,021	2044	9,999	42,059	3.8
95–96	0.200502	8,977	1800	8,077	32,060	3.6
96–97	0.216376	7,177	1553	6,401	23,983	3.3
97–98	0.233093	5,624	1311	4,969	17,582	3.1
98–99	0.250634	4,313	1081	3,773	12,614	2.9
99–100	0.268969	3,232	869	2,797	8,841	2.7
100+	1.00000	2,363	2363	6,044	6,044	2.6

Arias E. United States life tables, 2003. National vital statistics reports; vol 54 no 14. Hyattsville, MD: National Center for Health Statistics. 2006.

100 years old; their data are simply grouped into a 100+ catch-all category. This is because there weren't enough people in this category to give us statistically reliable year-by-year data. The good news, at least in the United States, is that this conclusion is changing and we should expect to see tables going out further in the near future.

The second column (labeled q_x) is the probability of dying during the corresponding year x. You can see that the year of birth ($x = 0$, the top row,) in the table is a relatively dangerous year. The probability of dying actually falls with age for a few years after birth, then starts rising again, but doesn't exceed the birth year probability until the 56th year. Being born is apparently a somewhat risky process: There are issues of birth defects leading to an unsustainable infant life, problems occurring during birth, and so on.

Note also that q_{100}, the catch-all probability that (if you reach age 100) you will certainly die at some age greater than 100, is 1.000, the certain event. The reasoning behind this should be pretty obvious.

The numbers in the second column are independent of each other, they are not cumulative. That is, the probability of you dying in your 20th year assumes that you have succeeded in reaching your 20th year.

The third column (labeled l_x) is the number of people (survivors) reaching the starting age for a given row (x) assuming that we began with 100,000 people. In order to celebrate your 10th birthday, you must have already celebrated your first, second, third, . . . , and ninth birthdays. These numbers are therefore cumulative and since the result at each year depends on the result for the previous year, the relationship is referred to as recursive. Mathematically, since $1 - q_x$ is the probability of *not* dying in year x, we have

$$l_{x+1} = l_x(1 - q_x)$$

Looking at Table 5.1, l_1 is easy to understand. All it says is that everyone who is born must be alive on the day that they're born. Since we're starting with a population of 100,000, l_1 must be 100,000. Again, let me emphasize that 100,000 is just a convenient number to start with so that we can track deaths and remaining population over the years. Now, as pointed out above, if q_1 is the probability of someone dying during their first year of life, then $1 - q_1$ must be the probability of someone *not* dying during this period—that is, making it to their first birthday. If we are starting with 100,000 people and the probability of each of them reaching their first birthday is

$$1 - q_1 = 1 - .006865 = 0.993135$$

then

$$l_1(1 - q_1) = 100,000(0.993135) = 99,313$$

of these people will be alive on their first birthday, and this is the value of l_2.

The rest of the l table is built following this procedure, with each result being used to calculate the next (again, this is called a *recursive* formula).

The fraction of people alive at the start of a given year of life (x) is just $l_x/100,000$. For example, the number of people reaching their 25th year is just .98219 or approximately 98.2% of the people born 25 years earlier. Now between their 25th and 26th years, people will die at a more-or-less uniform rate. Just to keep things simple, however, let's assume that 98.2% of the original group of people is the average number of people alive during their 25th year. The conclusions reached in the calculations below will therefore be not perfectly accurate, but the points to be made will be (hopefully) made a bit clearer.

BIRTH RATES AND POPULATION STABILITY

Before going through the rest of the columns in the Life Table, let's see what we can learn from what we have so far. In order for there to be 100,000 25-year-old people alive, the starting group must have had 100,000/.982 = 100,813 babies born, or 1.00813 more people born than reaching their 25th birthday. Assuming that half of the population is women and that they all arranged to give birth during, and only during, their 25th year, each woman would have had to give birth to 2(1.00813) = 2.01626 babies. This is of course a somewhat nonsensical perspective. Women who have children do not have them all in their 25th year, and they certainly do not average twins or more per birth. Births occur, statistically, over a wide range of (womens') lives. The point to be made here is that it is not enough for each woman, on the average, to give birth to a total of two children, or even for each couple, on the average, to give birth to one child per person, for the population to be stable. The l_x column of the Life Table decreases with each year. Other than by immigration of people into the United States, there is no way to replace people who die each year; and in order for the population to be stable, the average number of births per woman must be greater than two.

Going back to the simple assumption of all births taking place in one year, what if I change that year from age 25 to age 40? Looking at the Life Table and repeating the calculation, we find that the average number of births per woman jumps from about 2.016 to about 2.074. This makes sense because some women have died between ages 25 and 39, and therefore a higher average birth rate per mother is needed to maintain the population. This effect has actually been happening in the United States because more and more women want to have both families and established careers and are consequently starting families later in life than did earlier generations. Advances in medical care that make this decision safe have certainly also influenced the decision.

The actual birth rate is of course based upon a weighted average of all the birth rate statistics. The bottom line, however, is still that an average birth rate of greater than two babies per mother, over the mother's life, is necessary to maintain a constant total population. If the birth rate is lower than this number, the population will continue to fall; if the birth rate is higher than this number, the population will continue to grow. In the former case, the population of people will eventually disappear. In the latter case, eventually the number of people will outpace the ability to feed these people and the Life Tables (due to changes in the q column) will change. Over the course of history you can see many, many factors influencing the Life Tables in addition to the simple availability of food: Wars, disease, religious beliefs and practices, and so on, all weigh in. As will be shown below, however, the ability to accurately predict the cost of a life insurance policy many years in advance depends on the stability of the Life Tables, or at least on a good picture of how they're going to change over the years. This of course is just a restatement of everything we've looked at so far: If the coin you're about to flip is continually being replaced with another coin with a new (and unknown) probability of landing on heads, you can't know what your betting odds should be to keep the game fair.

LIFE TABLES, AGAIN

Returning to the Life Tables, I should point out that all of the information in the table is based on the first column (age, or x) and on the second column, q_x. Everything else is calculated from these first two columns, except for subtle details of how things are changing during the course of the year. The fourth column, d_x, is the number of people dying at a given age. That is,

$$d_x = l_x - l_{x-1}$$

This is a fairly straightforward calculation to understand. If 100,000 people are born and only 99,313 reach their first birthday, then $100,000 - 99,313 = 687$ people must have died between birth and their first birthday.

The fifth column, L_x, is the number of person-years lived between ages x and $x + 1$. This is a calculation we'd need if, for example, we knew that the "average person" of age x needs 2 quarts of water a day and wanted to know much water we would need to supply a large group for a year.

If we assume that people die at a uniform rate, then we would expect we get L_x by averaging l_x and l_{x+1}, that is,

$$L_x = \frac{l_x + l_{x+1}}{2}$$

Trying this formula for $x = 40$, we get

$$L_{40} = \frac{96,444 + 96,249}{2} = 96,346$$

Comparing this to the value for L_{40} in the table (96,346), it looks like we've got things right. However, if we try the same calculation for $x = 0$, we get 99,657 as compared to the table value of 99,394. Clearly, there is a problem here: Our simple calculation yields a number that is much too high. This discrepancy is explained by some details that we don't have in front of us: For people about 40 years old, the death rate is fairly uniformly distributed over the year. This means that the average number of people alive during the year is identically the number of people alive at the middle of the year. For $x = 0$ we are looking at the first year of life, including the birth process and the hours immediately after birth. Since the birth process and the next few hours are inherently dangerous, we should expect a higher-than-average death rate early on in the year (for $x = 0$). This will give us an actual L_0 lower than that predicted by the simple formula above. Note that for $x = 1, 2$, and so on, the simple formula again does a good job.

The sixth column, T_x, is the total number of person-years lived above age x. These numbers are calculated by starting at the bottom of the table and working up, that is

$$T_x = T_{x+1} + L_x$$

Starting at the bottom of the table, T_{100} must equal L_{100} because $x = 100$ is our final "catch-all" row in the table. What this assertion is saying is that the total number of people-years lived between age 100 and all higher ages (L_{100}) must be the same as the total number of people-years lived above age 100 (T_{100}). Then, the total number of people-years lived above age 99 (T_{99}) must just be the sum of the total number of people years lived *above* age 100 (T_{100}) and the total number of people years lived between age 99 and age 100 (L_{99}). Follow this recursive procedure all the way up to the top of the table and you'll have all of the T values.

T_x is useful for predicting how much resources our population will need for the rest of their lives. If, for example, we want to give every 65-year-old person a certain pension and continue it until they die, how much money should we anticipate needing, on the average, per person? At age 65 we have 82,668 people (l_{65}) and we are predicting 1,524,128 person-years to come (T_{65}), so we should anticipate having to pay these people for an average of $T_{65}/l_{65} = 18.4$ years.

This last calculation is shown in the seventh and last column, $e_x =$ expectation of life at age x:

$$e_x = \frac{T_x}{l_x}$$

When you're at some age x, e_x tells you how many more years you can expect to live. Your expected age at death is then just $x + e_x$.

Looking at the e_x column, we see that at birth the average life expectancy was 77.5 years, whereas at age 65 it was $65 + 18.4 = 83.4$ years. This seems very different from the situation we have when flipping a coin. The coin has no memory. If you just had the greatest run of luck ever recorded—that is, you just flipped a coin 100 times and got 100 heads—the expected result of the next 10 flips is still 5 heads and 5 tails. Why should your life expectancy behave differently—why should it change as you age?

The answer to this seeming paradox is that *your* personal life expectancy hasn't changed at all. What has happened is that the population base for the statistics has changed: Some babies have succumbed to birth defects and early childhood diseases, some teenagers have gotten drunk and driven into trees, and so on. By the time you reach 65 years of age, all the people who, for whatever reason, were to die at a younger age have already died and those that are left form a group with a longer life expectancy. This is also called *biased sampling* and is logically not that different from the "marbles on a board" example in Chapter 1.

Let's look at a few other interesting numbers we can glean from these tables. The average age of the population is just the average of all ages weighted by the number of people alive at that age l_x (weighted averages were discussed in Chapter 2). Again, we realize that we are introducing a small error by assuming that l_x is the average number of people alive during the year x and also by the lumping together of everyone 100 or more years old, but as a good approximation, we calculate

$$\text{average age} = \frac{\sum_{x=0}^{100} x l_x}{l_x} = 40.8 \text{ years}$$

Without showing the calculation, the standard deviation of this number is 24.7 years. The standard deviation is more than half of the average. In other words, don't expect most of the people you run into on the street every day to be about 41 years old. Also, note that two standard deviations is 49.4 years. This is a case where the distribution is clearly not normal, since it's not quite possible to have people less than 0 years old. If we want a confidence interval about the average age, we have to look carefully at the distribution, we cannot simply count the number of standard deviations.

To calculate the median age, we simply start summing up the l_x numbers from $x = 0$ down and from $x = 100$ up, and watch for the point where these sums cross. In other words, at what age do we have as many people younger than that age as we do people older than that age? Without showing the table entries, this turns out to be 39.5 years old. This is approximately the same as the average age. The mode age is defined as the age at which we have the largest population. This is of course $x = 0$ years, a number totally unlike the mean or the median. Again, non-normal distributions can give very non-normal calculation results.

PREMIUMS

Now let's look into the calculation of life insurance premiums. Remember that I'm ignoring costs, profit margins, and so on, and just treating life insurance premium calculations as a fair game gambling exercise. If you're just at your 65th birthday and you want to buy a year's worth of term insurance, then by looking at the q_x value (column 2) for $x = 65$, you see that your premium should just be $.015005 for every $1.00 of insurance. This is because if your probability of dying during the year between your 65th and 66th birthdays is .015005, then this is exactly the amount that the insurance company must charge for every dollar they would pay if you "won" the bet in order to make the game fair. A $100,000 one-year term life insurance policy should cost $1500.50. Remember that in practice, however, in addition to administrative costs and profits, your life insurance company will also be looking at tables that are much more specific to your age, sex, current health, and so on, than our table—so don't expect to actually see this number.

Let's improve the above calculation a bit. l_x = .015005 isn't really the probability of dying at any time in your 65th year. The actual value must start (right at your 65th birthday) a bit higher than this and end (right before your 66th birthday) a bit lower than this so that an actual l_x curve calculated day-by-day rather than a year at a time is a smooth curve. This is because it just doesn't make sense for the value to take a sudden jump the day of your 66th birthday. Let's assume that this value is an average value for the year and by working with this average we'll get a reasonable, if not as accurate as possible, result. Using the same reasoning, we have to look at the cost of this term policy. When you write a check and hand it to your insurance broker, you are loaning the insurance company money that they don't have to repay until the day you die (or not at all if you don't die during the term of the policy). Assuming that you will die during your 65th year, the average time that the insurance company will be holding your money is 1/2 year. This is not the same as a coin-flip game where money changes hands every minute or so. The fair game calculation must now take consider the *present value* of $100,000 returned to you 1/2 year after you gave it to the insurance company.

Present value is a simple but important financial calculation: Using your best estimate of the interest that you could earn on your money, how much money would you have to put into a savings account today so that you could withdraw $100,000 1/2 year from now? Just to keep things simple, let's assume that the bank compounds interest twice a year. This means that if we assume that the annual interest rate is 5%, the bank will give you 2.5% interest on your money twice a year. Your cost for the insurance should therefore be reduced by the same factor, that is,

$$\text{premium} = \frac{1500.50}{1.025} = 1463.88$$

What about the cost of buying a $100,000 policy on your 65th birthday that will cover you for the rest of your life? One way to solve this problem is to calculate the present value (on your 65th birthday) of one-year term policies for each year from your 65th to your "100+" years and then add up the cost of all of these policies. There is an important difference between this proposed policy and the one-year term policy discussed above. In the case of the term policy, if your probability of dying during the term of the policy is, say, .01, then there's a .99 probability that the insurance company will get to keep your money and never pay out. In the case of a policy that covers you from a start date until your death, there is no doubt that at some time the insurance company will have to pay. This latter condition is reflected in the q_x column of the life table for age 100+ as a probability of 1.00, the certain event. As the old saying reflects, "you can't get out of this alive."

Consider Table 5.2. The first four columns are copied directly out of the Life Table shown earlier, but only the rows for $x = 65$ and higher are shown. The fourth column (d_x) shows the number dying during each year, based upon the original 100,000 people born. In order to turn this into a PDF, we must *normalize* this column—that is, add up all of the entries and then divide each entry by this sum. What the normalization procedure does is to scale all the numbers so that they add up to 1. This result is a PDF for the probability of your dying at a given age, assuming that you've reached your 65th birthday. The results of this calculation are shown in column 5. Note that the first entry in this column is the same as the entry in the second column (l). This is because both of these entries represent the same number—the probability of you dying during your 65th year, assuming that you are alive on your 65th birthday. All of the other entries in this column differ from the entries in the second column because the second column is the probability of dying at age x assuming that you have reached your xth birthday, whereas the fifth column is the probability of you dying at age x looking forward from your 65th birthday. For example, look at the bottom row, $x = 100+$: The second column has probability of 1. This means that it is very certain that once you have reached your 100th birthday, you are going to die sometime after your 100th birthday. The fifth column has a probability of ~.03. This means that looking forward from your 65th birthday, there is approximately a 3% probability that you will live past your 100th birthday.

Figure 5.1 shows this (5th) column. Remember that this is the probability of your dying at a given age, looking from the vantage point of your 65th birthday. There are a couple of funny-looking features to this figure. First, there is a little bump in the curve at age 83. This is just the "statistics of statistics,"—that is, the fact that data from real sources almost never produce expected values exactly. Second, there is the jump at age 100. This is because the age = 100 information is really all ages from 100 up lumped into a final catch-all row in the table. If we had real data for ages 100, 101, 102, and so on, the curve would decline gracefully to 0.

TABLE 5.2. Cost per Year of $100,000 Policy Starting at Age 65

Age	q_x	l_x	d_x	Probability of Dying	Time to Mid-Year	Cost Divider per 1/2 Year	Cost for $100,000 Policy
65–66	0.015005	82,668	1240	0.015005	0.5	1.025	1463.88
66–67	0.016240	81,428	1322	0.015996	1.5	1.077	1485.42
67–68	0.017837	80,105	1429	0.017284	2.5	1.131	1527.69
68–69	0.019265	78,676	1516	0.018335	3.5	1.189	1542.45
69–70	0.021071	77,161	1626	0.019667	4.5	1.249	1574.78
70–71	0.023226	75,535	1754	0.021222	5.5	1.312	1617.45
71–72	0.024702	73,780	1823	0.022046	6.5	1.379	1599.27
72–73	0.027419	71,958	1973	0.023867	7.5	1.448	1647.91
73–74	0.029698	69,985	2078	0.025142	8.5	1.521	1652.29
74–75	0.032349	67,906	2197	0.026572	9.5	1.599	1662.17
75–76	0.035767	65,710	2350	0.02843	10.5	1.680	1692.66
76–77	0.039145	63,360	2480	0.030002	11.5	1.765	1700.21
77–78	0.042748	60,879	2602	0.031481	12.5	1.854	1698.05
78–79	0.046289	58,277	2698	0.032632	13.5	1.948	1675.31
79–80	0.051067	55,579	2838	0.034333	14.5	2.046	1677.73
80–81	0.056846	52,741	2998	0.036267	15.5	2.150	1686.84
81–82	0.061856	49,743	3077	0.03722	16.5	2.259	1647.72
82–83	0.067173	46,666	3135	0.037919	17.5	2.373	1597.81
83–84	0.077268	43,531	3364	0.040688	18.5	2.493	1631.86
84–85	0.079159	40,168	3180	0.038463	19.5	2.620	1468.28
85–86	0.086601	36,988	3203	0.038748	20.5	2.752	1407.89
86–87	0.094663	33,785	3198	0.038687	21.5	2.891	1337.95
87–88	0.103381	30,587	3162	0.03825	22.5	3.038	1259.11
88–89	0.112791	27,425	3093	0.037418	23.5	3.192	1172.35
89–90	0.122926	24,331	2991	0.03618	24.5	3.353	1078.96
90–91	0.133819	21,340	2856	0.034545	25.5	3.523	980.54
91–92	0.145499	18,485	2689	0.032534	26.5	3.701	878.96
92–93	0.157990	15,795	2495	0.030187	27.5	3.889	776.25
93–94	0.171312	13,300	2278	0.027561	28.5	4.086	674.58
94–95	0.185481	11,021	2044	0.024728	29.5	4.292	576.08
95–96	0.200502	8,977	1800	0.021773	30.5	4.510	482.79
96–97	0.216376	7,177	1553	0.018785	31.5	4.738	396.48
97–98	0.233093	5,624	1311	0.015858	32.5	4.978	318.57
98–99	0.250634	4,313	1081	0.013077	33.5	5.230	250.04
99–100	0.268969	3,232	869	0.010516	34.5	5.495	191.39
100+	1.00000	2,363	2363	0.028582	35.5	5.773	495.11

Continuing, column 6 shows the number of years from your 65th birthday to the middle of year *x*. The argument here is the same as was given above in the term insurance example: We are assuming that the (insurance company's) bank compounds interest every half year and that the deaths all occur at mid-

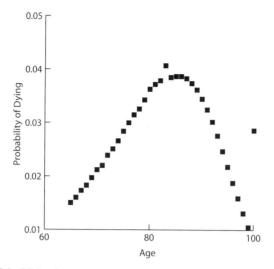

Figure 5.1. PDF of age at death looking forward from 65th birthday.

year. The seventh column shows the correction factor on the $100,000 payoff based on the number of years that the insurance company gets to keep the money. This again is just an extension of the term insurance calculation:

$$\frac{\text{cost}}{\$100,000} = (1+.025)^{2n}$$

where n is the number of 1/2 years from your 65th birthday to your death and we are again assuming 5% annual interest compounded twice a year.

The last column is the product of the probability of your dying in a given year (column 5), the present-value cost correction factor (column 6), and $100,000. This column shows the cost, at your 65th birthday, of insuring you every year thenceforth. The top entry, $x = 65$, is of course the same result as for the one-year term insurance example above.

The total cost of the policy is just the sum of all of the numbers in the last column, $44,526.82. Is this a reasonable number? The longer you wait before buying a life insurance policy that covers you for the rest of your life, the more it will cost you. If, for example, you wait until your 100th birthday, then according to our tables the $100,000 policy will cost $100,000 (minus a small bit for present value correction). If you were to buy the policy at, say, age 45, this same $100,000 policy would only cost $21,859.75. Be careful before concluding that you are saving a lot of money by buying the policy at an earlier age. You must calculate the present value of both choices at the same date before you can compare actual costs. Buying the policy at the earlier age of course insures your family against your accidental or other untimely death, which is a consideration that just might be more important than the relative costs of the policies.

SOCIAL SECURITY—SOONER OR LATER?

Next, I'll look at a slightly different calculation that ultimately will need the life tables: The Social Security Administration tells me that if I start collecting Social Security payments at age 66, I'll get (approximately) $2000 a month for the rest of my life, but if I wait until I'm 70 years old to start collecting payments, I'll get (approximately) $2700 for the rest of my life. What should I do?

I'll look at some extreme cases that don't require much calculation first. Assume that I'm now just coming up to my 66th birthday and I have to make a decision about when to start collecting Social Security payments. If I'm unfortunate enough to be suffering from some terminal disease and know that I probably won't last until I'm 70 years old, then my choice is obvious; start collecting the money now and enjoy it while I can. On the other hand, if every member of my family for six generations back has lived well into their nineties and right now I'm looking pretty good myself, then I probably should wait until I'm 70 years old to start collecting—the higher payments for all of the years past my 70th birthday will more than compensate for the four years' wait.

But what about the intermediate case(s)? I have no terminal disease, and my family life expectancy history tends to look like the one in the life tables. What should I do? I'd like to look at this problem in three different ways. I'll be conservative about interest rates—let's say that 4% is a reasonable safe investment (CDs or something like that).

The first way to look at this is to just start taking the money immediately (at age 66) and putting it in the bank. Just to keep the tables small, I'll only compound interest and receive funds twice a year. That is, I'll assume that the Social Security checks are $2000 × 6 = $12,000 twice a year and that the bank gives me 2% interest on my balance twice a year. The twice a year check if had I waited until age 70 would have been $2700 × 6 = $16,200. If I were to take the difference, $16,200 − $12,000 = $4200, from my savings account twice a year after I reach age 70 and add it to my ($12,000) check, my financial life wouldn't know which choice I had made—until my savings account ran out.

Table 5.3 shows my savings account balance over the years. The savings account runs out just past halfway through my 86th year. I'll hold a discussion of what I can conclude from this table until I've gone through the other ways of looking at this problem.

The second way of comparing the two choices is to calculate the present value of both choices at the same date. In this case a reasonable choice of this date is my 66th birthday, since that's when I have to make my decision. Since I don't know when I'm going to die, all I can do is make a list of the present values of each choice versus a possible year of death.

Figure 5.2 shows both present values versus my possible age of death. While the choice of starting payments at age 66 looks better if I die soon, the two

TABLE 5.3. Savings Account Balance if I Start Social Security Payments at Age 66–70

Age	Withdrawal	Balance	Age	Withdrawal	Balance
66.0	$0.00	$12,000.00	76.5	$4,200.00	$68,810.00
66.5	$0.00	$24,240.00	77.0	$4,200.00	$65,986.20
67.0	$0.00	$36,724.80	77.5	$4,200.00	$63,105.93
67.5	$0.00	$49,459.30	78.0	$4,200.00	$60,168.05
68.0	$0.00	$62,448.48	78.5	$4,200.00	$57,171.41
68.5	$0.00	$75,697.45	79.0	$4,200.00	$54,114.84
69.0	$0.00	$89,211.40	79.5	$4,200.00	$50,997.13
69.5	$0.00	$102,995.63	80.0	$4,200.00	$47,817.08
70.0	$4,200.00	$100,855.54	80.5	$4,200.00	$44,573.42
70.5	$4,200.00	$98,672.65	81.0	$4,200.00	$41,264.89
71.0	$4,200.00	$96,446.11	81.5	$4,200.00	$37,890.18
71.5	$4,200.00	$94,175.03	82.0	$4,200.00	$34,447.99
72.0	$4,200.00	$91,858.53	82.5	$4,200.00	$30,936.95
72.5	$4,200.00	$89,495.70	83.0	$4,200.00	$27,355.69
73.0	$4,200.00	$87,085.61	83.5	$4,200.00	$23,702.80
73.5	$4,200.00	$84,627.32	84.0	$4,200.00	$19,976.86
74.0	$4,200.00	$82,119.87	84.5	$4,200.00	$16,176.39
74.5	$4,200.00	$79,562.27	85.0	$4,200.00	$12,299.92
75.0	$4,200.00	$76,953.51	85.5	$4,200.00	$8,345.92
75.5	$4,200.00	$74,292.58	86.0	$4,200.00	$4,312.84
76.0	$4,200.00	$71,578.44	86.5	$4,200.00	$199.09

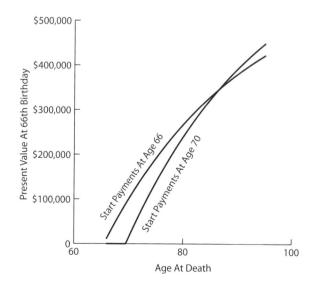

Figure 5.2. Present value of Social Security payments versus starting age.

TABLE 5.4. Weighted Current Values of Social Security Payments

Age	Current Value of $12,000 Payment	Current Value of $16,200 Payment	d_x	Probability of Dying	Cumulative Probability of Dying	Probability of Seeing Birthday	Weighted Current Value of $12,000 Payment	Weighted Current Value of $16,200 Payment
66	$12,000.00	$0.00	1322	0.008213	0.008213	1.0000	$12,000.00	$0.00
66.5	$11,764.71	$0.00	1376	0.008544	0.016757	0.9918	$11,901.44	$0.00
67	$11,534.03	$0.00	1429	0.008874	0.025631	0.9832	$11,567.57	$0.00
67.5	$11,307.87	$0.00	1472	0.009144	0.034775	0.9744	$11,018.04	$0.00
68	$11,086.15	$0.00	1516	0.009414	0.044188	0.9652	$10,700.63	$0.00
68.5	$10,868.77	$0.00	1571	0.009756	0.053944	0.9558	$10,388.50	$0.00
69	$10,655.66	$0.00	1626	0.010097	0.064041	0.9461	$10,080.85	$0.00
69.5	$10,446.72	$0.00	1690	0.010497	0.074538	0.9360	$9,777.70	$0.00
70	$10,241.88	$13,826.54	1754	0.010896	0.085434	0.9255	$9,478.47	$12,795.94
70.5	$10,041.06	$13,555.44	1788	0.011108	0.096542	0.9146	$9,183.21	$12,397.34
71	$9,844.18	13,289.64	1823	0.011319	0.107861	0.9035	$8,893.81	$12,006.64
71.5	$9,651.16	13,029.06	1898	0.011786	0.119647	0.8921	$8,610.18	$11,623.74
72	$9,461.92	12,773.59	1973	0.012254	0.131901	0.8804	$8,329.83	$11,245.27
72.5	$9,276.39	12,523.13	2026	0.012581	0.144482	0.8681	$8,052.83	$10,871.32
73	$9,094.50	12,277.58	2078	0.012908	0.15739	0.8555	$7,780.51	$10,503.69
73.5	$8,916.18	12,036.84	2138	0.013276	0.170666	0.8426	$7,512.86	$10,142.36
74	$8,741.35	11,800.82	2197	0.013643	0.184309	0.8293	$7,249.50	$9,786.82
74.5	$8,569.95	11,569.43	2273	0.01412	0.198429	0.8157	$6,990.43	$9,437.09
75	$8,401.91	11,342.58	2350	0.014597	0.213025	0.8016	$6,734.73	$9,091.89
75.5	$8,237.17	11,120.18	2415	0.015	0.228025	0.7870	$6,482.45	$8,751.30
76	$8,075.66	10,902.14	2480	0.015404	0.243429	0.7720	$6,234.20	$8,416.17

76.5	$7,917.31	10,688.37	2541	0.015784	0.259213	0.7566	$5,990.01	$8,086.51
77	$7,762.07	10,478.79	2602	0.016163	0.275376	0.7408	$5,750.04	$7,762.56
77.5	$7,609.87	10,273.33	2650	0.016459	0.291834	0.7246	$5,514.30	$7,444.30
78	$7,460.66	10,071.89	2698	0.016754	0.308588	0.7082	$5,283.38	$7,132.57
78.5	$7,314.37	9,874.40	2768	0.017191	0.325779	0.6914	$5,057.24	$6,827.28
79	$7,170.95	9,680.78	2838	0.017628	0.343407	0.6742	$4,834.81	$6,526.99
79.5	$7,030.34	9,490.97	2918	0.018124	0.361531	0.6566	$4,616.08	$6,231.71
80	$6,892.49	9,304.87	2998	0.01862	0.380151	0.6385	$4,400.65	$5,940.87
80.5	$6,757.35	9,122.42	3038	0.018865	0.399016	0.6198	$4,188.54	$5,654.52
81	$6,624.85	8,943.55	3077	0.01911	0.418126	0.6010	$3,981.43	$5,374.93
81.5	$6,494.95	8,768.18	3106	0.019289	0.437415	0.5819	$3,779.25	$5,101.98
82	$6,367.60	8,596.26	3135	0.019469	0.456883	0.5626	$3,582.32	$4,836.13
82.5	$6,242.74	8,427.71	3249	0.02018	0.477063	0.5431	$3,390.54	$4,577.23
83	$6,120.34	8,262.46	3364	0.02089	0.497953	0.5229	$3,200.55	$4,320.74
83.5	$6,000.33	8,100.45	3272	0.020319	0.518272	0.5020	$3,012.45	$4,066.80
84	$5,882.68	7,941.62	3180	0.019748	0.53802	0.4817	$2,833.85	$3,825.70
84.5	$5,767.33	7,785.90	3191	0.019821	0.557841	0.4620	$2,664.39	$3,596.93
85	$5,654.25	7,633.23	3203	0.019894	0.577735	0.4422	$2,500.08	$3,375.10
85.5	$5,543.38	7,483.56	3201	0.019879	0.597614	0.4223	$2,340.77	$3,160.05
86	$5,434.68	7,336.82	3198	0.019863	0.617476	0.4024	$2,186.84	$2,952.24
86.5	$5,328.12	7,192.97	3180	0.019751	0.637227	0.3825	$2,038.13	$2,751.48
87	$5,223.65	7,051.93	3162	0.019639	0.656866	0.3628	$1,895.00	$2,558.25
87.5	$5,121.23	6,913.65	3128	0.019425	0.676291	0.3431	$1,757.27	$2,372.31
88	$5,020.81	6,778.09	3093	0.019211	0.695502	0.3237	$1,625.28	$2,194.13
88.5	$4,922.36	6,645.19	3042	0.018894	0.714396	0.3045	$1,498.85	$2,023.44
89	$4,825.84	6,514.89	2991	0.018576	0.732972	0.2856	$1,378.28	$1,860.68
89.5	$4,731.22	6,387.15	2923	0.018156	0.751128	0.2670	$1,263.37	$1,705.55

TABLE 5.4. Continued

Age	Current Value of $12,000 Payment	Current Value of $16,200 Payment	d_x	Probability of Dying	Cumulative Probability of Dying	Probability of Seeing Birthday	Weighted Current Value of $12,000 Payment	Weighted Current Value of $16,200 Payment
90	$4,638.45	6,261.91	2856	0.017736	0.768864	0.2489	$1,154.38	$1,558.41
90.5	$4,547.50	6,139.13	2773	0.01722	0.786084	0.2311	$1,051.09	$1,418.97
91	$4,458.33	6,018.75	2689	0.016704	0.802788	0.2139	$953.71	$1,287.51
91.5	$4,370.92	5,900.74	2592	0.016101	0.818889	0.1972	$862.00	$1,163.70
92	$4,285.21	5,785.04	2495	0.015499	0.834388	0.1811	$776.10	$1,047.73
92.5	$4,201.19	5,671.60	2387	0.014825	0.849212	0.1656	$695.77	$939.29
93	$4,118.81	5,560.40	2278	0.01415	0.863363	0.1508	$621.07	$838.44
93.5	$4,038.05	5,451.37	2161	0.013423	0.876786	0.1366	$551.75	$744.86
94	$3,958.87	5,344.48	2044	0.012696	0.889482	0.1232	$487.79	$658.52
94.5	$3,881.25	5,239.69	1922	0.011937	0.901419	0.1105	$428.95	$579.08
95	$3,805.15	5,136.95	1800	0.011179	0.912598	0.0986	$375.11	$506.40
95.5	$3,730.53	5,036.22	1676	0.010412	0.92301	0.0874	$326.06	$440.18
96	$3,657.39	$4,937.47	1553	0.009645	0.932655	0.0770	$281.58	$380.14
96.5	$3,585.67	$4,840.66	1432	0.008893	0.941548	0.0673	$241.48	$325.99
97	$3,515.37	$4,745.74	1311	0.008142	0.94969	0.0585	$205.48	$277.40
97.5	$3,446.44	$4,652.69	1196	0.007428	0.957118	0.0503	$173.39	$234.08
98	$3,378.86	$4,561.46	1081	0.006714	0.963832	0.0429	$144.89	$195.60
98.5	$3,312.61	$4,472.02	975	0.006057	0.969889	0.0362	$119.81	$161.74
99	$3,247.66	$4,384.33	869	0.005399	0.975288	0.0301	$97.79	$132.02
99.5	$3,183.98	$4,298.37	1616	0.010037	0.985325	0.0247	$78.68	$106.22
100	$3,121.54	$4,214.09	2363	0.014675	1	0.0147	$45.81	$61.84

curves cross and the choice of starting payments at age 70 looks better if I live a long life. The two curves cross just past half way through my 86th year. This isn't surprising, because this calculation really is the same as the first calculation above, it's just another way of looking at things.

These calculations give me some interesting numbers to look at and ponder, but they don't help me to make a decision. I can only make the actual correct decision if I know exactly when I'm going to die. Since I don't know this date (fortunately, because I don't think I could live happily if I knew this date many years in advance), I have to fall back on the next-best thing, the statistics. That is, I'll repeat my present value calculations but I'll weigh them by the information I get from the life tables.

Table 5.4 looks fairly complex, but it's really just a combination of things that have been presented before. The first column is age, in 1/2 year increments. The second and third columns are the current value of each of the $12,000 payments starting at age 66 and the $16,200 payments starting at age 70, respectively. In the table I refer to them as *current* values just to avoid confusing them with my previous usage of *present* value—which was the sum of all of the current values up to a given age.

Column 4 is d_x, the number of people dying at age x, taken from the Life Table, Figure 5.1. Column 5 is the PDF of the probability of dying at age x, calculated by normalizing column 4, as described above. Column 6 is the cumulative probability of dying, calculated from column 5 as described in Chapter 2.

Column 7 is the probability of seeing your birthday at age x. This is calculated, row by row, by just taking 1 − column 6. This starts at 1 because the table starts at age 66; and if I've reached my 66th birthday, it's a certainty that I'll be alive on my 66th birthday (sort of a nonsensical but correct statement).

Column 7 is column 4, the current value of the $12,000 payments, weighted by the probability that I'll be alive to collect this payment. Column 8 is the same for the $16,200 payments. The weighted averages I want are just the sum of each of these two columns, which turn out to be $289,204 for starting to collect my Social Security payments at age 66 and $272,389 for starting to collect my Social Security payments at age 70.

These two numbers are so close to each other that there doesn't seem to be a "winner." They are the same to within about ±3%. Playing around with the interest rate a bit could make them even closer (or further apart). Also, the Life Table I'm using is now a few years old and is for the total population—not the population about to start collecting Social Security. The bottom line seems to be that it really doesn't matter which choice I make.

CHAPTER 6

BINOMIAL PROBABILITIES

If you had 10 coins (are you getting sick of thinking about flipping coins yet?) that are numbered $1, 2, \ldots, 10$ and you wanted to know the probability of flipping a head on the first coin, a tail on the second coin, and so on, this would not be a difficult problem. Actually, this isn't *a* problem, it's 10 independent problems lumped together into one sentence. The probability of a given result with each coin is 0.5, so the probability of all 10 results is simply $(1/2)^{10} = 1/1024 \approx 0.001$, with the $\approx$ sign meaning "is approximately equal to." But what if you're interested in a slightly different problem, say of getting exactly 6 heads and 4 tails out of 10 flips, and you don't care which of the flips gives you heads and which gives you tails, so long as when you're done, there were 6 heads and 4 tails?

One way of solving this problem is to write down every possible result of flipping 10 coins and then to count the number of ways that you could get 6 heads and 4 tails. The probability would just be this number of ways multiplied by the probability of getting any one of them—or, equivalently, this number of ways divided by 1024. I actually followed this procedure in Chapter 1 when I wanted to see how many ways I could get a total of 6 when I rolled a pair of dice (Table 1.2). In the case of the pair of dice, there were only 36 combinations, so the writing-down exercise was a little annoying but not that terrible. For 10 coins there are 1024 combinations, so the writing-down exercise would take a couple of hours and really wouldn't be fun. If I got nasty and asked for

Probably Not: Future Prediction Using Probability and Statistical Inference,
by Lawrence N. Dworsky.
Copyright © 2008 John Wiley & Sons, Inc.

the probability of getting 15 heads and 5 tails when I flip 20 coins, then there are over 1 million combinations and writing them all down is clearly out of the question.

THE BINOMIAL PROBABILITY FORMULA

The answer to the need to calculate these probabilities without listing all of the possible outcomes is called the *binomial probability* formula. It is *binomial* in the sense that each of the repeated events (the coin flip, the die toss) has two possible outcomes. These outcomes have probabilities p and q, where q is $1 - p$, or the probability of *not p*. This is not as fancy an idea as it first appears; if p is the probability of flipping a head, then q is the probability of not flipping a head—that is, of flipping a tail. If p is the probability of rolling a 5 with a die ($p = 1/6$), then q is the probability of rolling anything but a 5 with the die ($q = 5/6$).

Suppose I want to flip a coin n times and I want to know the probability of getting k heads. I know that $0 \leq k \leq n$, which is a fancy way of saying that k can get as small as 0 (no heads) or as large as n (all heads). The probability of getting k heads is

$$\text{probability of } k \text{ heads} = p^k$$

The probability of getting everything else tails is then

$$\text{probability of "the rest of them" tails} = q^{n-k}$$

This last expression makes sense, but you just have to take it a piece at a time. Remember that if the probability of getting a head is p, then the probability of getting a tail is q. This seems trivial for a coin when $p = q = 0.5$, but if we had a "funny" coin that lands on heads 90% of the time, then $p = 0.9$ and $q = 0.1$. If I am flipping n coins and I want k of them to be heads, then $n - k$ of them must be tails.

The probability of getting k heads out of n coin flips when the probability of a head is p is therefore

$$(\text{number of ways})p^k q^{(n-k)}$$

where "number of ways" means the number of ways that we can achieve k heads when we flip n coins.

Before looking into a formula for calculating "number of ways," consider a couple of examples based on situations where it's easy to count these "number(s) of ways." Unfortunately, in the cases that you can easily count the number of ways, you usually don't need the whole binomial probability formula in the first place; but just to solidify the use of the terms in the formula, I'll go through a couple of these cases anyway.

The best examples to start with are the same coin flip examples as above, but with a small enough group of coins that I can easily do the counting.

Example 1. What's the probability of getting exactly 2 heads when I flip 5 coins? In this example $n = 5$ (the number of coins), $k = 2$ (the number of heads), and $p = 0.5$ Therefore,

$$p^k q^{n-k} = (0.5)^2 (0.5)^3 = (0.25)(0.125) \approx 0.0313$$

Table 6.1 shows all (32) possible combinations of flipping 5 coins. Looking forward, one of the benefits of developing the binomial probability formula is that I won't have to write out (and you won't have to stare at) one of these interminable tables again. Looking down the table, I see exactly 2 heads in 10 (of the 32) rows. The probability of getting exactly 2 heads when flipping 5 coins is therefore $(10)(.0313) \approx 0.313$.

The probability of getting exactly 1 head when flipping 5 coins is $(5)(.0313) \approx 0.157$, and so on.

Finishing the derivation of the binomial probability formula requires a discussion of counting, which in turn requires discussions of permutations, combinations, and factorials of numbers. If you're not interested, you can skip this; just remember that I'm ultimately calculating the *number of ways* that a given goal can be accomplished in a situation.

TABLE 6.1. All Possible Results of Flipping a Coin 5 Times

Coin					Coin				
1	2	3	4	5	1	2	3	4	5
H	H	H	H	H	T	H	H	H	H
H	H	H	H	T	T	H	H	H	T
H	H	H	T	H	T	H	H	T	H
H	H	H	T	T	T	H	H	T	T
H	H	T	H	H	T	H	T	H	H
H	H	T	H	T	T	H	T	H	T
H	H	T	T	H	T	H	T	T	H
H	H	T	T	T	T	H	T	T	T
H	T	H	H	H	T	T	H	H	H
H	T	H	H	T	T	T	H	H	T
H	T	H	T	H	T	T	H	T	H
H	T	H	T	T	T	T	H	T	T
H	T	T	H	H	T	T	T	H	H
H	T	T	H	T	T	T	T	H	T
H	T	T	T	H	T	T	T	T	H
H	T	T	T	T	T	T	T	T	T

Suppose that we have 4 objects, labeled A, B, C, and D, and we want to see how many ways we can order them, such as ABCD, ADCB, GDCA, and so on.

For the first letter, we have 4 choices (A, B, C, or D). For the second letter, we only have 3 choices—we already used one of the choices for the first letter. For the third letter we have 2 choices, and for the last letter we have only 1 choice left. This means we have (4)(3)(2)(1) = 24 choices. I promised both you and myself that I'd try to stop writing out long tables, so you'll have to write this one out yourself if you want to see all 24 choices.

The calculation (4)(3)(2)(1) is called 4 *factorial*, the notation for which is 4!. In general, the number of ways that we can order n objects is $n!$. Spreadsheet programs and scientific calculators all have a factorial function available, so this calculation isn't necessarily tedious. It does, however, get out of hand very quickly: 4! = 24, 5! = 120, 6! = 720, and so on. From this short list of examples you might have spotted a fundamental principle of factorials: $n! = n[(n - 1)!]$. This is another example of a *recursive* relationship.

PERMUTATIONS AND COMBINATIONS

What if we want to order 6 items, but only look at, say, 4 of them at a time? In other words, if I have a box with 6 items in it, how many ways can I order any 4 of them? The solution to this is just an extension of the factorial calculation: For the first item I have a choice of 6, for the second item a choice of 5, for the third item a choice of 4, and for the fourth item a choice of 3. Putting all of this together, the number of ways I can order 6 items, 4 at a time is (6)(5)(4)(3).

This looks like a "piece" of a factorial calculation. Realizing that I can write (6)(5)(4)(3) as

$$(6)(5)(4)(3) = \frac{(6)(5)(4)(3)(2)(1)}{(2)(1)} = \frac{6!}{2!}$$

I see that I can always write the number of ways that I can order n items, k at a time, as

$$P_{n,k} = \frac{n!}{(n-k)!}$$

I snuck in a new notation here. This formula is known as the *permutation* of n things taken k at a time.

We're almost where we want to go. There is a slight difference between $P_{n,k}$ (the number of ways we can order n things k at a time) and the number of ways a given goal can be accomplished. The difference is a consequence of the fact that in the former case order counts (ABCD is not the same as BACD), whereas in the latter case order does *not* count (ABCD is the same as BACD, is the same as DCAB, etc.) If you're interested in how many ways you can get 4 heads out of 6 flips of a coin, you don't care if you get HHHHTT or TTHHHH or THHHHT or . . . The permutation formula is giving us too large a number!

The number of ways we can order k items is $k!$. Therefore, if we take $P_{n,k}$ and divide it by $k!$, we'll have reduced the number of ways of ordering n items k at a time by the number of ways we can (re)arrange the k items. This is just what we are looking for—the number of ways our goal can be accomplished.

This relationship is known as the *combination* of n things taken k at a time and is written

$$C_{n,k} = \frac{P_{n,k}}{k!} = \frac{n!}{(n-k)!k!}$$

Occasionally, instead of $C_{n,k}$ you'll see $\binom{n}{k}$. It's just a different notation, it means the same thing.

$C_{n,k}$ is easy to calculate in principle, and for small values of n it's easy to calculate in practice. Again, spreadsheet programs and scientific calculators will have this function available for you. As an example, the combination of 10 things taken 6 at a time is

$$C_{10,6} = \frac{10!}{(6!)(4!)}$$

If you want to work this out, please don't calculate 10!, 6!, and 4! (unless you're looking for help falling asleep). Take advantage of the fact that

$$\frac{10!}{6!} = \frac{(10)(9)(8)(7)(6)(5)(4)(3)(2)(1)}{(6)(5)(4)(3)(2)(1)} = (10)(9)(8)(7)$$

So that

$$C_{10,6} = \frac{10!}{(6!)(4!)} = \frac{(10)(9)(8)(7)}{(4)(3)(2)(1)} = 210$$

Now let's put all of this together. The probability of getting 6 heads out of 10 flips of a coin (or a flip of 10 coins) is

$$C_{10,6}(0.5)^6(0.5)^4 = (210)(.015625)(.0625) \approx .205$$

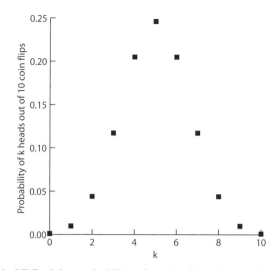

Figure 6.1. PDF of the probability of getting k heads out of 10 coin flips.

What about the probability of getting 4 heads out of 10 flips? Since the probability of a single head = the probability of a single tail = 0.5, then the probability of getting 4 heads out of 10 flips should be the same as the probability of getting 4 tails out of 10 flips, which of course is the same as the probability of getting 6 heads out of 10 flips.

Suppose you want the probability of getting *at least* 6 heads when you flip 10 coins? You just add up all of the individual probabilities—the probability of getting 6 heads (out of 10 flips), 7 heads, 8 heads, 9 heads, and 10 heads when you flip 10 coins.

Figure 6.1 shows the probability of getting k heads out of 10 flips; k can take on values from 0 to 10 ($0 \leq k \leq 10$). Just in case the question occurred to you, all 11 probabilities add up to 1, as they should.

LARGE NUMBER APPROXIMATIONS

Figure 6.1 looks suspiciously like (points on) a normal curve. The mean is 5. This makes sense—if I flip 10 coins, each with a probability of 0.5 of coming up heads, then I would expect the mean to be $(10)(0.5) = 5.0$. Remember that for a normal distribution, the mean is the same as the expected value. In general, we should expect the mean to be the product of the number of coin flips (or whatever we're doing), n, and the probability of success at each even, p:

$$\mu = np$$

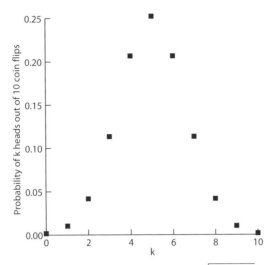

Figure 6.2. Normal distribution with $\mu = np$, $\sigma = \sqrt{np(1-p)}$ approximation to Figure 6.1.

Figure 6.2 shows a normal distribution with the mean as shown above and the standard deviation equal to

$$\sigma = \sqrt{np(1-p)}$$

Deriving this last relationship is a little messy, so I'll ask you to just *trust me* (or better yet, look up the derivation).

These last two formulas are the exact formulas for the mean and standard deviation of a binomial distribution.

These two PDFs (binomial and normal) are so much alike that if I were to overlay them on the same graph, you'd have trouble realizing that there are two sets of data being presented. In retrospect, the normal distribution is an excellent approximation to the binomial distribution for any but small values of n (say, $n < 5$), and the approximation gets better as n gets larger, so long as p isn't very small or very large. This shouldn't be a surprise, because in Chapter 3 I showed that every distribution starts looking normal when you add more and more cases of it (the distribution) together.

Using these formulas for the mean and the standard deviation of a normal distribution approximation to the binomial distribution can save a lot of time when looking at problems with large values of n. It is necessary, of course, to be able to quickly calculate values of the normal distribution—but again, this is available on spreadsheets and scientific calculators (as is the binomial distribution itself).

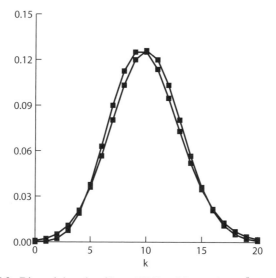

Figure 6.3. Binomial and uniform PDFs with $n = 1 \times 10^9, p = 1 \times 10^{-8}$.

Figure 6.3 shows a situation where the binomial and the normal distributions look pretty similar, but they're not exact. They're close enough that I won't bother to label which curve is which, the point being that the uniform distribution now only looks like a *fair* approximation to the binomial approximation. The numbers that were used to generate Figure 6.3 are $n = 1$ billion $= 1,000,000,000$ ($= 1 \times 10^9$ in scientific notation) and $p = 10/(1 \text{ billion}) = .00000001$ ($= 1 \times 10^{-8}$ in scientific notation). One billion is equal to 1000 million (and that's the way it's described in Great Britain). The expected value of the probability is then ($np =$) 10. This is a pretty unlikely event: only 10 "hits" out of 1 billion tries! Since the normal PDF looks pretty much like the binomial PDF, the standard deviation calculated using the formula above is accurate. Using this formula, we obtain

$$\sigma = \sqrt{np(1-p)} = \sqrt{\mu(1-p)} = \sqrt{10(.9999999)} \approx \sqrt{10} \approx 3.16$$

For a normal distribution, the 95% confidence interval is $\pm 2\sigma$, which in this case is 6.32 above and 6.32 below the mean. Does this make any sense? The number of successes, or "hits" must be an integer. What does 6.32 mean?

Remember that the normal distribution is a continuous function approximation to the binomial distribution. The binomial distribution is a discrete PDF—it's only defined for integer values of k. If we want to take advantage of our normal PDF approximation without too much extra work, then we'll just say that the mean plus or minus 6 is a confidence interval of "just under" 95% and leave well enough alone for now.

THE POISSON DISTRIBUTION

Now let's look at a *really* improbable event. Leave n at a billion, but drop p to one in a billion (1×10^{-9}). The binomial and normal PDFs for these values are shown in Figure 6.4. In Figure 6.3 the dots (data points) are connected just to make it easier for your eye to follow which data points belong to which curve. These are discrete distributions, only valid at the data points (values of k), not in between.

Figure 6.4 shows that in the extreme case of a very rare occurrence out of very, very many trials, the binomial distribution doesn't look at all like the normal distribution. For that matter in the latter example the normal distribution would extend its 95% confidence interval into negative values of k, which makes no sense whatsoever.

The Poisson distribution is defined only for non-negative integer values of k. The theory of the Poisson distribution is beyond the scope of this book. The important properties of the Poisson distribution are as follows:

1. The Poisson distribution is a discrete distribution, valid only for integer arguments.
2. The Poisson distribution is never negative.
3. When the Poisson distribution is approximating a binomial distribution, the mean value (usually called λ) is given by $\lambda = np$.
4. The standard deviation is $\sigma = \sqrt{\lambda}$.

For those of you who want to see the actual function, the value of the Poisson distribution for k occurrences with a mean value of λ is

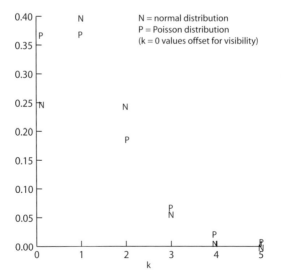

Figure 6.4. Binomial and uniform PDFs with $n = 1 \times 10^{9}$, $p = 1 \times 10^{-9}$.

$$P_{k,\lambda} = \frac{e^{-\lambda}\lambda^k}{k!}$$

The Poisson distribution is a good approximation to the binomial distribution for $n > \sim 200$ and either p or q very close to 1 (which of course makes q or p, respectively, very close to 0).

The Poisson distribution was discovered by Simeon-Denis Poisson and published in the mid-1800s. He showed that the Poisson distribution describes the probability of k occurrences of some event if these events occur at a known rate (λ) and the occurrence of each event is independent of the occurrence of the last event.

As an example of using the Poisson distribution for this application, suppose that every day you take a walk through the park at lunchtime. You walk down a quiet lane and count the number of cars that pass you by. Over the course of a year the average number of cars that pass you by is 10 per hour. The Poisson distribution (Figure 6.5) will predict the PDF for the number of cars that pass you by each day.

Figure 6.5 looks symmetric, and it's not unlike a normal distribution. However, note that the mean occurs at $k = 10$ while the value of the PDF at $k = 9$ is the same as the value at the mean. This PDF is asymmetric. For $k > 5$, however, the 95% confidence interval may be approximated reasonably well by assuming the PDF is normal. For smaller values of k the Poisson PDF is notably asymmetric and you must go to the formulas (spreadsheet, scientific, or financial calculator) to do your calculations.

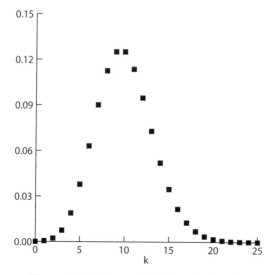

Figure 6.5. Poisson distribution for $\lambda = 10$.

DISEASE CLUSTERS

When people with some tie between them such as family relations, friendships, residence neighborhood, or drug (legal or illegal) usage start developing unusual diseases, there's reason to be suspicious. If the average incidence in the United States of some particular disease is about 1 in 100,000 per year, then the expected incidence in a city of 1 million population of this disease is 10 per year. If the incidence in our city (of 1 million people) is 15 this year, we'd work through the probabilities and conclude that we should keep our eye on things but not jump to any conclusions. If 6 of these 15 know each other, or are related to each other, or all live within 1/4 mile of the Acme Chemical Plant, or something like that, however, then it's not a coincidence. Or is it?

Historically, important public health discoveries have studies that began with observations such as these. Two examples are the long-term danger of asbestos inhalation and the initial understanding of the nature of AIDS. Unfortunately, things sometimes go the other way. The courts decided that silicone breast implants caused serious illnesses and awarded enough money to plaintiffs that the implant manufacturer, Dow Chemical, declared bankruptcy because it couldn't pay all the settlements. I doubt that the fact that several large studies subsequently showed that there was no evidence supporting the tie between these breast implants and the diseases claimed caused any guilt-ridden plaintiffs to mail the money back.

I need to discuss the inevitability of a highly unlikely event given enough opportunities for the event to occur. This has already come up in the discussion of slot machine winnings, but it's worth starting from the beginning and working through this because this subject is, quite reasonably, very important to many people. There are two topics that come to mind:

First: Probability of an unlikely event in a large population. "Deadly cosmic rays are randomly hitting the earth and striking 100,000 people nightly." What should the incidence rate (expected value and standard deviation) look like in your city of 1 million inhabitants?

Second: We have a disease (or cosmic rays as above) that affect one person in 100,000, with all the statistics attached to this. We have a new factor (watching too much reality TV?) that increases your chance of contracting the disease or succumbing to the cosmic rays, or … , which we think is causing 5 extra cases of the disease per 100,000. What should the expected number of cases and the standard deviation look like?

CLUSTERS

Suppose you live in a city of 1/4 million people. If there's a noncommunicable disease (I'll look at contagion in a later chapter)—such as a brain cancer—that afflicts, say, 1 person in 5000 each year, then we would expect the average incidence of this disease in your city to be (using the binomial probability formulas)

$$\mu = np = (250,000)(.0002) = 50$$

cases per year.

Still using the binomial probability formulas, the standard deviation should be

$$\sigma = \sqrt{\mu(1-p)} = \sqrt{50(1-.0002)} = 7.07$$

As an aside, note that to a very good approximation, we obtain

$$\sigma = \sqrt{\mu(1-p)} = \sqrt{50(1-.0002)} \approx \sqrt{50} = \sqrt{\mu}$$

which is the relationship we see when dealing with a Poisson distribution. In this case the normal distribution is also a good approximation. Without showing the graph,[1] I'll assert that the binomial formula for the probability of some number of cases overlaps the normal distribution curve for the same μ and σ so well that it's hard to tell that two separate curves have been drawn.

The 95% confidence interval is therefore essentially between 43 and 57 cases. Any number of cases falling outside these numbers is usually deemed to be "statistically significant."

Let's assume that this year your town has 70 cases of this disease. This is certainly well above expected statistical variations and deserves further study.

An odd but not unexpected human trait is that 30 cases this year would also deserve further study, newspaper articles, and television specials—maybe there's something very beneficial going on in this town with regard to incidence of this disease. Somehow this doesn't seem to happen too often.

Anyway, again from the binomial formulas, the probability of getting 70 cases this year is just 0.0010. The probability of getting 70 or more cases is 0.0026. Now, let's assume that there are 1000 towns around the country more or less like yours. Using these same formulas yet one more time, we find that the expected number of towns with 70 or more cases of this disease is approximately 2.5. It's pretty certain that there will be at least one town out there with 70 or more cases of this disease, even though when you look at just one of these towns at a time, it's statistically very unlikely that this can happen unless there's a special cause.

I could really beat this discussion tc death if I chose to. I could add statistics about town sizes, perhaps statistics about distance to the nearest nuclear power plant, the number of cars driving around, and the annual chocolate ice

---

[1] At this point in this book I feel that I've shown the normal distribution curve so many times that if I just list the mean and the standard deviation, you can picture what I'm describing.

cream consumption. The bottom line conclusion would still be "Be very careful about assigning a cause to what might be a truly random event." Look for asbestos, look for toxic industrial waste, and look for nuclear materials, but also be open to just having plain dumb old bad luck. Never forget to do the numbers—fully.

CHAPTER 7

PSEUDORANDOM NUMBERS AND MONTE CARLO SIMULATIONS

Suppose I want to generate a list of uniformly distributed random numbers. These could be integers between 0 and 10, or any real number between 0.0 and 1.0 or some other requirement. Your first question is probably something like "Why would you want to do this?" The answer is that this enables me to write computer programs that perform *Monte Carlo* simulations. I'll go into this in detail and with examples later on in this chapter, but the short answer is that a Monte Carlo simulation lets me perform, on the computer, a simulation of any situation involving random numbers that I can describe clearly and allows me to look at possible results. If I repeat this procedure many, many times, then I get to see the statistical variations in these results. As a simple example, I'll drag out my coin again. If I had a list of randomly generated zeroes and ones, then I could assign a zero to a coin flip heads and assign a one to a coin flip tails. By counting the number of zeros and ones and looking at the ratio, I could conclude that the probabilities of heads and tails is 0.5 each. If I looked at, say, 1 million lists of zeros and ones, with each list containing 100 of these numbers, then I could calculate the standard deviation I'd expect in a 100-coin flip game. I could look at the distribution of heads (or tails) in this list of 1 million games and convince myself that this PDF is (or isn't) Gaussian. I could then calculate the standard deviation (of the mean) and then confidence intervals on the PDF. In other words, I could do on the

Probably Not: Future Prediction Using Probability and Statistical Inference, by Lawrence N. Dworsky.

computer what I can't do in real life, namely, set up fairly involved situations and let them "do their thing" many, many times and then study the statistics of the results.

PSEUDORANDOM NUMBERS

There are what are called *hardware random number generators*. These are machines that measure random noise or similar physical phenomena that we believe to be truly random (with some known PDF) and convert these measurements to strings of numbers. These machines, however, are costly and run slowly (here we are defining slowly as compared to the blazing calculating speeds of even small personal computers). We prefer to use something inexpensive and fast.

Computer scientists and mathematicians have learned how to generate what are called *pseudorandom* numbers using computer programs called, reasonably enough, *pseudorandom number generators* (PSNGs).

A computer program that is fed the same input data will produce the same results every time it is run. Numbers coming out of a PSNG are therefore not random. Every time you start a PSNG with the same input parameters (call *seeds*), it will produce the same string of numbers. This sequence of numbers will successfully approximate a sequence of random numbers based on how well it achieves a sophisticated set of criteria. These criteria include the probabilities of successive digits occurring, the mean and sigma of a large sequence, whether some properties of the generator can be deduced from studying a string of numbers that this generator produced, and so on.

The mathematics of pseudorandom number sequences and trying to refute the randomness of a proposed sequence is a very sophisticated field, and I'll hardly be scratching the surface in this discussion.

PSNGs must depend on a fixed formula, or algorithm, to generate their results. Sooner or later, every PSNG will repeat. Another requirement for a good PSNG is therefore that it doesn't repeat until it has generated many more numbers than we need for our problem at hand.

As a very simple example, for a string of pseudorandom integers between 1 and 10, I'll propose the list 1, 2, 3, 4, 5, 6, 7, 8, 9, 10, 1, 2, 3, 4, 5, 6, 7, 8, 9, 10, 1, 2, 3, 4, . . .

This list looks pretty good if I calculate the mean and the standard deviation. However, since this list repeats after every tenth number, if it's usable at all, it will only be usable when we need 10 or fewer numbers. Using only the first 10 numbers gives us such a short list that statistics tend to be poor, but let me continue on just to demonstrate the use of the evaluation criteria. We would immediately flunk even the short list of the first 10 numbers from the above list because, after the first few numbers, it's pretty easy to guess the next number. This is easily circumvented—we just scramble the list to, say, 4, 7, 3, 5, 8, 2, 10, 1, 6, 9.

Even after scrambling, this list is still suspicious—there are no repeats. While a sequence of 10 real numbers between 1 and 10 would almost never have any repeats, a list of integers in this range would almost always have at least one repeat. Later on in this chapter, I'll show how to use a Monte Carlo simulation to calculate the probability of this happening.

THE MIDDLE SQUARE PSNG

An early computer-based PSNG is known as the *middle square* method. Suppose I want a random number between 0 and 1, with 4 digits of resolution (how to extend this to any number of digits will be obvious). Start with a 4-digit "seed" number as shown in Table 7.1. I'll start with 3456 as my seed, as seen in the upper left corner of the table. When I square a 4-digit number, I get an 8-digit number—the second column in the table. I then extract the inner 4 digits of the 8-digit number (column 3) and divide it by 10,000 (column 4) to get my number between 0 and 1. Finally, I use these new (inner) 4 digits as the seed for my next random number (row 2, column 1) and repeat the process.

Table 7.1 shows only the first 15 numbers of the sequence. How does it do as a pseudorandom distribution? A uniform distribution between 0 and 1 should have a mean of 0.500. These 15 numbers have a mean of 0.476. Sigma should be 0.289, and this group has a sigma of 0.308. The smallest number is 0.0625, the largest number is 0.974. So far, this group looks pretty good.

TABLE 7.1. Sample Random Number List Generated by the Middle Square Procedure

Seed	Seed Squared	Inner Four Digits	Divide by 10,000
3456	11943936	9439	0.9439
9439	89094721	947	0.0947
947	00896809	896	0.8968
8968	80425024	4250	0.425
4250	18062500	625	0.0625
625	00390625	3906	0.3906
3906	15256836	2568	0.2568
2568	06594624	5946	0.5946
5946	35354916	3549	0.3549
3549	12595401	5954	0.5954
5954	35450116	4501	0.4501
4501	20259001	2590	0.259
2590	06708100	7081	0.7081
7081	50140561	1405	0.1405
1405	1974025	9740	0.974

It seems as if I've got the situation under control. This simple procedure gives me what I wanted. Unfortunately, this procedure has some pathological problems. Table 7.2 shows what happens if the number 3317 comes up. The sequence decays to zero and never leaves it again. Table 7.3 shows what happens if the number 1600 shows up. In this case the sequence gets *stuck* in a pattern that repeats every fourth iteration. This doesn't look very random at all.

This procedure could be "patched" by having the computer program that's generating these numbers look for these pathological situations and jump away from them by perturbing the seed, but then we'd have to carefully examine the consequences of these fixes to the statistics of the numbers generated. In general, this is really not a good way to proceed.

TABLE 7.2. What Sometimes Goes Wrong with the Middle Square Procedure

Seed	Seed Squared	Inner Four Digits	Divide by 10,000
3317	11002489	0024	0.0024
0024	00000576	0005	0.0005
0005	00000025	0000	0.0000
0005	00000025	0000	0.0000
0005	00000025	0000	0.0000
0005	00000025	0000	0.0000
0005	00000025	0000	0.0000

TABLE 7.3. Something Else that Can Go Wrong with the Middle Square Procedure

Seed	Seed Squared	Inner Four Digits	Divide by 10,000
1600	02560000	5600	0.5600
5600	31360000	3600	0.3600
3600	12960000	9600	0.9600
9600	92160000	1600	0.1600
1600	2560000	5600	0.5600
5600	31360000	3600	0.3600
3600	12960000	9600	0.9600
9600	92160000	1600	0.1600
1600	2560000	5600	0.5600
5600	31360000	3600	0.3600
3600	12960000	9600	0.9600
9600	92160000	1600	0.1600
1600	2560000	5600	0.5600
5600	31360000	3600	0.3600

THE LINEAR CONGRUENTIAL PSNG

The PSNG procedure used in PCs today is called a *linear congruential* generator. The formula is very simple. There are three constants, A, B, and M. With
a starting value V_j, the next value, V_{j+1}, is

$$V_{j+1} = (AV_j + B) \bmod M$$

where mod is shorthand for a modulus operation. What that means is that you
take the number in front of mod, divide it by M, and only look at the remainder. A few examples of this are 5 mod 4 = 1, 7 mod 5 = 2, 6 mod 3 = 0, and
so on.

Table 7.4 shows this formula used for the values $A = 11$, $B = 7$, and $M = 23$. The starting value is 4. Since this formula will generate numbers between
0 and 22, a random number between 0 and 1 is generated by dividing the
results of this formula by M.

**TABLE 7.4. Sample Random Number List Generated by the Linear Congruential
Generator, Using $A = 11$, $B = 7$, $M = 23$**

V	$AV + B$	$(AV + B)$ mod 23	Random Number
4	51	5	0.217391
5	62	16	0.695652
16	183	22	0.956522
22	249	19	0.826087
19	216	9	0.391304
9	106	14	0.608696
14	161	0	0
0	7	7	0.304348
7	84	15	0.652174
15	172	11	0.478261
11	128	13	0.565217
13	150	12	0.521739
12	139	1	0.043478
1	18	18	0.782609
18	205	21	0.913043
21	238	8	0.347826
8	95	3	0.130435
3	40	17	0.73913
17	194	10	0.434783
10	117	2	0.086957
2	29	6	0.26087
6	73	4	0.173913
4	51	5	0.217391
5	62	16	0.695652

A continuous uniform distribution of random numbers between 0 and 1 would have a mean of 0.500 and a sigma of 0.2887 (it takes a little integral calculus to calculate this latter number, so just trust me here). The list in Table 7.4 has a mean of 0.460 and a sigma of 0.2889. Not bad. The lowest number in the list is 0, and the highest number is 0.9565. There is no (or certainly no obvious) pattern to the spacing between the numbers, or any other discernible characteristic. This is a pretty darned good pseudorandom list.

The *linear congruential* generation of pseudorandom sequences will repeat *at most* after m numbers. Whether or not it actually makes it to m numbers before repeating depends on the choices for Z and I. Software in PCs today use an algorithm with M greater than 10 million, in some cases up into billions, and appropriately chosen values for Z and I.

Once I have a PSNG that I trust that gives me uniformly distributed numbers between 0 and 1, I can easily extend this to a PSNG that gives me uniformly distributed numbers between any lower and upper limits. If V_1 is such a number between 0 and 1, and I want numbers between a lower limit a and an upper limit b, then V_2 given by

$$V_2 = a + (b-a)V_1$$

is exactly what I want. To see how this formula works, just remember that V_1 will be between 0 and 1. If $V_1 = 0$, then $V_2 = a$. If $V_1 = 1$, then $V_2 = a + (b - a) = b$. Any number between 0 and 1 will come out somewhere between a and b.

Next, I'd like to look at the generating pseudorandom sequences with other than uniform distributions. There are many ways of doing this and I'm only going to discuss two of them. These two aren't optimum in any sense, but they work and they give excellent insight into the processes involved. There are much more modern, more efficient ways to do both jobs.

Both of these procedures assume the availability of a PSNG such as described above—that is, a PSNG that generates a uniform (pseudo)random sequence between 0 and 1.

A NORMAL DISTRIBUTION GENERATOR

The first example is a way to generate a normal distribution with a parameters μ and σ. I'll take advantage of the Central Limit Theorem, which tells me that adding up random numbers from any distribution will ultimately give me a normal distribution. My candidate in this case for *any* distribution is the uniform distribution between 0 and 1. Figure 7.1 shows this uniform distribution, labeled $n = 1$. This is followed by graphs of the distributions I get when I add this uniform distribution to another uniform distribution, then add the results to another uniform distribution, and so on—exactly as was described in Chapter 3. I keep going until I've combined 12 uniform distributions ($n = 12$ in the figure). Along the way, I scaled each distribution vertically so

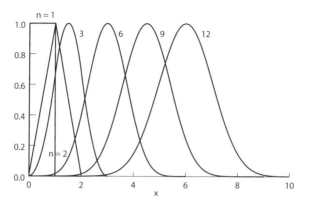

Figure 7.1. Several cases of up to 12 uniform distributions between 0 and 1 combined.

that the peak is at 1. This isn't necessary, but it makes the composite figure easier to comprehend.

It's pretty easy to see what happens to the mean value as I combine distributions. Remember that combining a distribution from 0 to 1 with another distribution from 0 to 1 gives me a distribution from 0 to 2, combining this with a distribution from 0 to 1 gives me a distribution from 0 to 3, and so on. Since these are symmetric distributions, the mean occurs exactly in the middle. Therefore a distribution from 0 to 6 has a mean of 3, and so on. Putting it in general terms, for n uniform distributions from 0 to 1 combined, the mean is at $n/2$.

At this point you might be wondering why 12 uniform distributions were chosen. The curve looks "pretty darned normal" after 6 distributions. On the other hand, why stop at 12? Figure 7.2 answers this question. σ and n of Figure 7.2 are referring to the distributions of Figure 7.1. When $n = 12$, σ is exactly 1. This is very convenient because it leads to the following procedure for generating normally distributed random variables:

1. Generate 12 uniformly distributed random variables between 0 and 1, x_i.

2. Add them up: $\sum_{i=1}^{12} x_i$

3. Subtract 6. This will bring the mean to 0: $\sum_{i=1}^{12} x_i - 6$

4. Multiply by the desired σ: $\sigma \left(\sum_{i=1}^{12} x_i - 6 \right)$

5. Add the desired mean. This correctly shifts the distribution:

$$\sigma \left(\sum_{i=1}^{12} x_i - 6 \right) + \mu$$

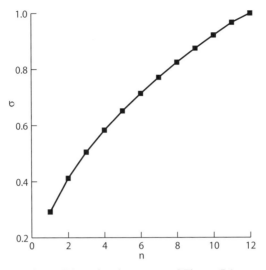

Figure 7.2. σ for the curves of Figure 7.1.

The results of this scheme are of course only as good as the PSNG, and it requires using the PSNG 12 times for every normally distributed number generator. Still, it works well and it's simple and clear.

AN ARBITRARY DISTRIBUTION GENERATOR

Next I'll present a procedure for generating random variables distributed arbitrarily. First, I need an example of an *arbitrary* distribution for an example. My only restriction is that the x values are only allowed over a finite range. That is, I must be able to specify specific starting and stopping values for x.

Figure 7.3 shows a pretty arbitrary PDF. I don't know of any real use for this thing or place where it occurs, I just made it up as an example. This is a continuous PDF, but the procedure to be described works just the same for a discrete PDF with points spaced along the lines of the figure.

The first thing I'll do is to scale this PDF so that the maximum value is 1. This is no longer a legitimate PDF, but right now I don't care about that. The only thing I want to make sure of when I stretch (or shrink, as the case may be) the PDF is that I keep the correct proportions.

Figure 7.4 shows the result of the scaling. I've also drawn a line across the top at the value 1. I'll explain why I did this very soon.

The first thing I do is to choose a random number between 2 and 6 using my uniform distribution random variable generator. Let's say that I came up with 3.500. (For a discrete PDF, I would randomly choose one of the allowed values of x.) Next, I choose a second random number between 0 and 1 using the same generator. Consider Figure 7.4. At $x = 3.500$, the value of the

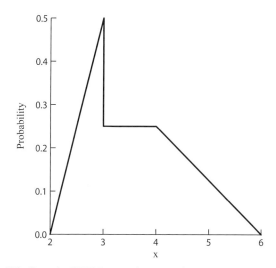

Figure 7.3. Sample PDF for random number sequence generation.

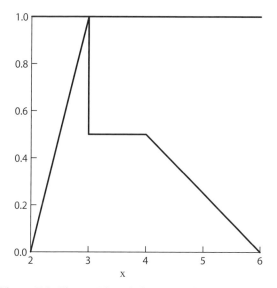

Figure 7.4. Figure 7.3 scaled to a maximum value of 1.

function is 0.5. If my second random number is below 0.5, then I take 3.500 as my result. If it's greater than 0.5, I start all over again.

Let's try this again: Choose a random number between 2 and 6 using the uniform distribution random variable generator (PSNG). Let's say that I came up with 5.00. Next, I choose a second random number between 0 and 1 using the same generator. Look at Figure 7.4 again. At $x = 5.00$, the value of the function is 0.25. If my second random number is below 0.25, then I take 3.500 as my result. If it's greater than 0.25, I start all over again.

Choosing 3.500 and choosing 5.00 from a uniform distribution between 2 and 6 are equally likely. However, choosing a number below 0.5 from a uniform distribution between 0 and 1 is twice as likely as choosing a number below 0.25 from this same distribution. If I created a long sequence of random numbers using this procedure, I would expect to see many more numbers from the regions where my distribution function is large (close to 1) than I would from where it is small. Fortunately, this is exactly what I'm looking for.

This procedure isn't very efficient in that it's not very *smart*. It doesn't really calculate anything, it just keeps trying and rejecting numbers that don't fit the criterion. It is very nice in that it's robust (it never gets into trouble) and it can be explained with absolutely no equations!

MONTE CARLO SIMULATIONS

Now I have the tools to start looking at Monte Carlo simulations. The question I posed earlier in this chapter was to find the probability that a list of 10 single-digit integers (i.e., the integers from 0 to 9) would have at least one repeat.

The Monte Carlo method is a brute force method. A mathematician colleague of mine once likened it to "putting a paper bag over your head and running down the road." This isn't far from the truth. The way the program will work will be to generate, using the computer's PSNG, a list of 10 integers between 0 and 9. Then the computer will count the number of repeated integers in the list and save this number. I'll have the computer repeat this procedure a million times or so, making sure that I never restart the PSNG at its initial seed values. Then I'll simply summarize the results.

Table 7.5 shows the first 20 lists of 10 single-digit integers that my program produced. The first list has two 0s (shown underlined). The second list has four 5s, two 8s, and two 9s. The third list has two 2s, two 8s, and three 9s. It's beginning to look like it's impossible to generate one of these lists without any repeats.

When I ran my program 1000 times, it did indeed tell me that 100% of the lists have at least one repeat. After 5000 iterations, my program found one list that had no repeats. That's 99.98% of the lists that had at least one repeat. After 10,000 iterations I had 99.95%, after 100,000 iterations I had 99.96%, and after 1 million and then 10 million iterations I had 99.96%. It looks like 99.96% is about the right answer. Note that I haven't proven this conclusion, I've just kept narrowing my confidence interval as I increased the number of iterations. A nonrigorous but pretty reasonable way of *testing* whether or not you've run enough iterations of a Monte Carlo simulation is to just keep increasing the number of iterations by large factors and watch for the answer to "settle down." In many cases the computer time involved can get quite long. Fortunately, computer time these days is a pretty inexpensive commodity.

TABLE 7.5. Twenty Lists of 10 Single-Digit Random Integers

List Number	List									
1	0	0	6	8	9	1	5	3	2	7
2	0	5	5	8	5	5	8	9	7	9
3	2	4	8	6	9	9	1	8	9	2
4	9	4	3	5	3	7	6	5	5	0
5	6	4	5	8	4	7	0	1	2	7
6	0	1	6	5	7	3	4	4	9	9
7	5	1	6	3	4	0	1	1	4	0
8	5	4	5	6	0	0	8	3	1	1
9	9	9	1	8	6	6	6	5	1	2
10	7	5	8	2	1	2	3	0	6	6
11	1	1	0	9	5	0	5	7	4	4
12	0	2	2	5	1	6	1	6	2	6
13	0	1	3	7	3	8	5	7	6	3
14	8	5	7	1	5	7	9	9	7	7
15	5	0	1	5	7	2	6	5	8	5
16	9	3	7	9	0	9	2	4	6	3
17	9	0	2	8	4	9	6	0	8	2
18	5	1	9	5	1	5	8	3	4	3
19	7	9	2	7	3	9	0	0	8	5
20	9	0	0	8	1	9	3	2	8	4

Reflecting back on the earlier discussion of simple pseudorandom number generation, it is not impossible for the list 0, 1, 2, 3, 4, 5, 6, 7, 8, 9, even with the numbers scrambled, to be a random list of single-digit integers, but it sure is unlikely!

Incidentally, this example could have been done analytically quite easily just by thinking "in reverse." That is, I ask the question, "What is the probability that no two integers in a group of 10 integers are the same?" For the first integer I have a choice of 10 out of 10; that is, I can pick any integer between 0 and 9. For the second integer I must choose one of the remaining 9 integers (out of 10). For the third integer I have to choose one of the remaining 8 integers out of 10. Writing this whole thing out, I get

$$\left(\frac{10}{10}\right)\left(\frac{9}{10}\right)\left(\frac{8}{10}\right)\left(\frac{7}{10}\right)\left(\frac{6}{10}\right)\left(\frac{5}{10}\right)\left(\frac{4}{10}\right)\left(\frac{3}{10}\right)\left(\frac{2}{10}\right)\left(\frac{1}{10}\right) = .000363$$

The probability of getting at least one repeat is therefore $1 - .000363 = .999634 \sim 99.96\%$.

Next, I'll look at a more challenging Monte Carlo Simulation example, one that can't be done better by simple calculations. Figure 7.5 shows a large rectangular box with several objects inside it. Imagine that this outside box is the wall of an ant farm. The dot in the center just indicates the location of the

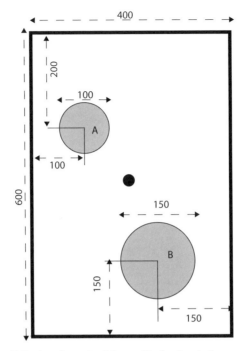

Figure 7.5. Ant farm for Monte Carlo simulation example.

center. I will be dropping ants, one at a time, onto the center. Unfortunately, my aim isn't too good. The ants will drop with an average location at the center of the box but with a normal distribution of distances from the center, with a standard deviation of 5. These are somewhat drunken ants; they wander around taking two-dimensional random walks. That is, every second they take a step, but there is an equal probability (1/4) of them stepping north, south, east, or west. If they bump into a wall, they bounce off. The step size is 3 units in the dimensions of the box.

The two gray circles represent holes in the bottom of the box. When an ant steps over the edge of a grey circle, it falls into the hole and is gone. Ultimately, all of the ants will fall into one of the holes.

I plan to drop 100 ants. I want to know how many of the ants will ultimately fall into each of the holes. Also, I'd like to know what the average and standard deviation of the number of steps an ant who falls into each of the holes takes.

How do I create a Monte Carlo simulation of this problem? First, I create what's called a *loop*. A computer program loop is a *wrapper* around a set of instructions. The computer executes all of the instructions, then goes back to the top of the list and starts over again. There is an index (a counter) that gets advanced each time the program repeats the loop, and typically there is a specified maximum number that this index can reach—that is, the number of

times that the program will execute the loop. This first loop index is the iteration number that the program is executing. This loop index is also a convenient
index for storing the results of the calculations in an array so that later on I
can calculate averages and sigmas.

Inside the iteration loop I create a second loop. This time I'm counting the
number of ants. Since I've decided that there are 100 ants, this inner loop will
execute 100 times. If my outer loop specifies 1000 iterations, by the time this
program is finished executing, 100,000 ants will have walked around in my ant
farm.

Inside this second, inner loop I am ready to drop an ant into the ant farm.
I do this by *requesting* a normally distributed random number with a mean of
0 and a sigma of 5. This gives me a radius the (statistically) correct distance
from the center of the box, but all I know now is the radius, which describes
a circle centered at the center of the box. Next I request a uniformly distributed random number between 0 and 360. This is the angle (in degrees) from
the horizontal that a line from the center to the radius would make. I now
have described a point; and using basic trigonometry, I get the location in
terms of horizontal and vertical (x and y) coordinates.

Now I am ready to start the ant's random walk. I start yet another loop.
This loop is different from the first two in that there is no index being counted.
This loop will continue endlessly until certain *exit criteria* are met. More of
this in a minute.

I request a uniformly distributed random number between 0 and 3.99999999,
then I strip away everything to the right of the decimal point. I now have one
of the four integers 0, 1, 2, or 3, with an equal probability (1/4) of any of them.
If the integer is 0, the ant moves to the right; if it's 1, the ant moves to the left;
if it's 2, and the ant moves up; and if it's 3, the ant moves down.

After each ant step, I have to check to see where the ant is. If it's at one
of the box edges, I just turn it around and send it back the other way. If it's
over one of the circles, I save the fact that it reached a circle and record which
circle it reached. Then I exit this loop and drop in another ant. If it's not over
one of the circles, I go back to the top of the loop (*loop back*), where I again
request a uniformly distributed random number between 0 and 3.9999,...,
and so on.

After 100 ants have fallen into one of the circles, I exit the second loop,
store my information, and *loop back* in the outer loop, thereby starting another
iteration and another 100 ants. When the program has completed the specified
number of iterations, it finishes by calculating averages and sigmas of the ants'
performance from the stored information.

Figure 7.6 is a *flow chart* of the computer programs operation. A flow chart
shows the logical flow of the program's operation, including loops and decision
points. This flow chart is a graphical depiction of the last few paragraphs. Not
shown in this flow chart are the instructions to (a) save the number of path
steps an ant took before falling into a hole and (b) record which hole the ant
fell into.

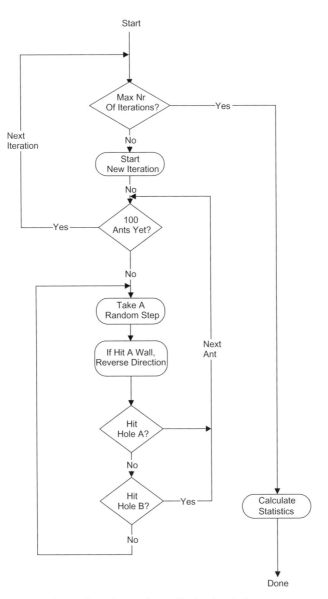

Figure 7.6. Flow chart of ant farm Monte Carlo simulation computer program.

Figures 7.7 and 7.8 are examples of ant walks. Note the huge variation in the number of steps taken: 820 in the first case and 9145 in the second.

As for the statistics, it turns out that 46% of the ants fall into hole A, 54% into hole B. The average path length (per iteration) is about 5800 to hole A and 4500 to hole B. This agrees with the probabilities of falling into the holes.

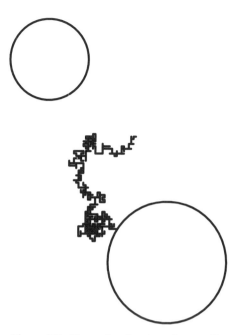

Figure 7.7. Example of ant random walk.

Since the ants are wandering randomly from near the center of the box, we should expect a higher probability that an ant will fall into a hole that's some combination of closer and larger. The standard deviations of the path lengths are 1300 for the walk to hole A and 920 for the walk to hole B. This at first seems odd: Why should the walks have different standard deviations? If I divide both of the standard deviations by the average path lengths, I get 22% and 20%, respectively. In other words, the longer the average path length to a hole, the higher the standard deviation in the path lengths to wander randomly to that hole—but the ratio of the standard deviation to the average path length stays pretty constant.

This last example was clearly a very contrived example. I'm not going to build an ant farm, and I don't really have a great interest in the results. As an example of a probability problem that's very straightforward to solve using a Monte Carlo simulation but impossible to solve by any analytic means, however, this ant farm is terrific. With very little more programming effort I could have changed the simple box to a maze, I could have added other holes, I could have given the ants a statistical distribution in their velocities, and so on. So long as I can coherently describe the attributes of the situation, I can create a Monte Carlo simulation of the situation.

Just as a cliff hanger, the comment about giving the ants a statistical velocity distribution will be a lead to talking about gas in a container and all the physics of the everyday world that this will let us describe.

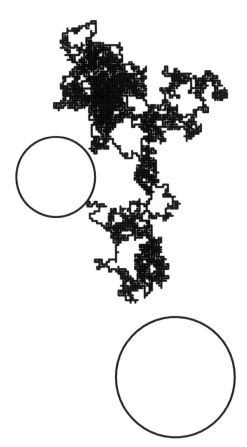

Figure 7.8. Example of ant random walk.

A LEAGUE OF OUR OWN

The last example of Monte Carlo simulations that I want to present represents a class of problems that have many solutions, some better than others. This example also is a bit contrived. It does, however, show the use of a very important concept called a Figure of Merit. This topic is really on the border of what's usually considered a topic in Probability.

In this type of problem, there is only a finite probability that we ever get to the "best" solution to our problem or even know if there is such a solution, but we usually are blissfully unaware of just what this probability is and only care that we can constantly better the odds of getting to the best solution (and can measure that we are indeed getting better).

The Acme Ephemeral Company has an in-house baseball league. There are five teams. Each team has a coach, a full roster of 9 players, and a few extra players scattered here and there among the teams. On these teams the pitchers are unique (they can only be pitchers) and the catchers are unique (they can only be catchers) but the infield players are fungible (each infield player can play any infield position) and the outfield players are similarly fungible. The league has been running for a few years now; several players have dropped out and several new players have joined. The problem is that the games are getting stale: Most of the players have been on the same team, with the same coach and many of the same players for most, if not all, of the years that the league has been in existence.

What I'd like to do is to scramble the players so as to minimize the "stale history." That is, I'd like each player to be placed with as many new teammates as possible, along with a new coach.

The first thing we have to realize is that there is no perfect solution. Each team, including the coach, is made up of at least 10 people. No matter how I distribute these 10 people among 5 teams, there is no way to keep some people from being with people they've been with before. Acknowledging that I can make things better but never perfect, I'd still like to proceed. Keep in mind that while it's easy to define what I mean by a perfect rearrangement, even though this is impossible to achieve, just what a "better" rearrangement that's short of perfect is will require some discussion.

Create a "trial" league by doing the following:

1. Randomly assign each coach to a team, one coach per team.
2. Randomly assign each pitcher to a team, one pitcher to a team. If there are excess pitchers, randomly assign them to a team, one pitcher to a team, until you run out of pitchers or there are two pitchers on every team. Repeat this procedure until you've run out of pitchers.
3. Repeat the pitcher process for the catchers.
4. Randomly assign all infielders to teams, at four infielders per team, and then follow the excess player procedure as you did for the pitchers.
5. Randomly assign all outfielders to teams, three outfielders per team, and then follow the excess player procedure as you did for the pitchers.

I now have a league that satisfies all the restrictions that are necessary to make all the teams functioning teams (they each have at least one coach, one pitcher, one catcher, four infielders, and three outfielders). Now I have to evaluate how well this league fits the rearranging goals.

I am going to create a number called a Figure of Merit (FOM). My definition will not be unique; I'll suggest some variants later. For now, just take it as a number whose starting value is 0.

I'm going to take every possible pair of players (for this exercise, the coach is considered to be a player) on these trial teams and then search all the team rosters for the past 3 years looking for the same pair of players together on a team in a given year. Each time I find a match, I'll add 1 to the FOM. When

I'm all done, I'll have a measure of how badly this trial league achieves my goal of a "total scramble." If I had not found any matches (this is impossible, as discussed above), the FOM would be 0.

Since this is my only trial league so far, this league and its FOM is the best I've ever created. I'll save the makeup of this league and the resulting FOM for the time being.

Now I repeat the process, that is, I randomly generate another league using the same procedures and then I calculate the FOM for this league. If the new FOM is the same or higher than my first FOM, I discard this latest league and start over again. When I get a new FOM that's lower than the FOM I've been saving, I discard the saved FOM and league and replace it with my latest FOM and league.

As I repeat this process many, many times, I will be driving the saved FOM down. In other words, I'm always working toward a league that more and more approaches my scrambling goal.

This process, as I've described it, never finishes. You never know if the next trial league won't be better (won't have a better FOM) than the one you're saving. It's easy to see that there are many, many millions of possible leagues that satisfy the basic constraints—that is, would-be leagues with fully functional teams. In principle I could write a computer program that methodically steps through every possible combination of players and calculates the FOM. If I lived long enough for this program to complete its run, I would know for sure that I had gotten the lowest possible FOM.

Figures of merit are typically used in problems that are too involved to solve analytically and where every design decision to make one thing a little better will make some other thing(s) a little worse. In many cases the FOM is defined as a measure of "goodness" rather than a measure of "badness" as I did above. In this latter case you work to maximize the FOM rather than minimize it. If you have bank savings and investments, you have already been working this kind of problem. Your FOM is the dollar value of your total savings and investments minus your debts. You regularly make decisions (buy this, sell that, etc.) to maximize your FOM.

When your FOM is the dollar value of your total savings and investments, the FOM is uniquely defined. In more complex problems, however, it is always possible to put weighting factors into the FOM definition. In my baseball league problem, for example, I might decide that it's most important to separate pitchers and catchers that have been together in the past. All I have to do is change the FOM "building" procedure so that I add, say, 2 rather than 1 to the FOM when I find a prior pitcher–catcher match. If it's really important, I could add 3, or 4, or whatever. Typically, there will be a table of the relative importance of the different contributing factors and the weights assigned to each of them which can be "played with" to see just what the computer program produces.

What I've been describing is certainly not a "hard" science or an exact mathematics. Assigning the weighting functions is an art that is done best by the person(s) most familiar with the real issues. And again, we're never sure when to stop the computer program and call it good enough.

When we're dealing with a problem that has continuous rather than discrete variables, the situation is usually handled somewhat differently. Suppose, for example, I'm designing a television. The marketing people tell me that customers want a large screen, but they also want the set to be as light as possible and of course as inexpensive as possible. Clearly, as the screen gets larger, the weight and price will go up. I would design an FOM that takes all of these factors into account and then sit down with the marketing folks to get their estimates on how much potential customers would pay for a larger screen, and so on, and translate this information into FOM weighting factors. The engineering design team will turn the relationships between the size, cost, and weight into equations. There are many mathematical techniques for formally searching the set of equations for optimum values to push the FOM down (assuming that I defined the FOM so that smaller is better). Some of these techniques are very smart in terms of looking for the absolute smallest (called the "global" minimum) value of the FOM without getting stuck in regions of "local" minima. None of them, however, can guarantee that they've found the best possible answer in a very complicated problem, and all of them require some oversight in choosing trial values of the variables.

Many of today's computer spreadsheet programs offer FOM minimizing (or maximizing) capability; the user defines the FOM in terms of relationships between the values in different cells of the spreadsheet. The step size guesses are usually suggested by the computer but can be overridden by the user. The relative weighting factors are provided by the user in building the formula for the spreadsheet cell that defines the FOM.

Some of the mathematical FOM search techniques available today are conceptually based upon natural processes. For example, there's an "annealing" technique that's modeled on the movement of molecules of metal as a liquid metal cools to its final (solid) state, there's a "genetic" technique that's modeled on evolutionary random mutations, and so on.

As I suggested above, these processes have as much art as they do science. Even assuming that we've found the best possible computer algorithms for optimizing our FOM, the results depend on the weighting factors put into the FOM, and therefore the overall value of the results is determined by the person(s) setting up the calculation.

CHAPTER 8

SOME GAMBLING GAMES IN DETAIL

By now you're probably sick of looking at coin flip games. Why do I keep talking about coin flip games? The answer is that the coin flip game is easy to describe and easy to model, so it's an excellent teaching platform. In this chapter I'll use tools such as the binomial PDF, confidence intervals, and Monte Carlo simulations to look at what happens when you take this simple game into a real gambling casino. You'll see why the house always comes out ahead and how they keep luring people by designing occasional big winnings into the system. I'll also talk about a TV game show that's a more involved version of the coin flip, and I'll show a statistical analysis of the Gantt Chart—a popular business planning tool. Lastly, I'll present a whole different approach to gambling, known as pari-mutuel betting.

THE BASIC COIN FLIP GAME

OK, the coin flip game again. Remember, the game is simple: Tails I give you a dollar, heads you give me a dollar. Assume a fair coin, that is, $P_{head} = P_{tail} = 1/2$. The expected value of your return on investment after a game is simply

$$E = (\text{winning amount})(P_{winning}) + (\text{losing amount})(P_{losing})$$
$$= (\$1)(0.5) + (-\$1)(0.5) = 0$$

Probably Not: Future Prediction Using Probability and Statistical Inference,
by Lawrence N. Dworsky.
Copyright © 2008 John Wiley & Sons, Inc.

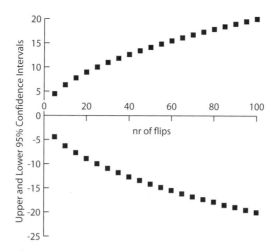

Figure 8.1. Ninety-five percent confidence interval of winnings versus number of coin flips.

The expected value after any number, n, of flips must similarly be exactly zero. This is what I am calling a *fair game*.

Figure 8.1 shows (the upper and lower bounds of) the 95% confidence interval of these expected winnings versus the number of coin flips.

The story here is pretty clear: It's the random walk "couched in sheep's clothing." As in the random walk example, the standard deviation, $\sigma = \sqrt{n}$, n being the number of coin flips. The expected earnings are 0, but the confidence interval gets wider and wider with increased number of coin flips. At $n = 50$, for example, you are 95% confident of going home with somewhere between 14 dollars won and 14 dollars lost. At $n = 100$, you can be 95% confident that you will go home with somewhere between 20 dollars won and 20 dollars lost.

Suppose you start out lucky and your first 2 flips yield heads. This is not improbable (1/4 probability). You now have 2 dollars of winnings in your pocket.

Going forward, the coin has no memory of your earnings (or, equivalently, of its own or of another coin's history). This means that you can expect earnings of 0 going forward, despite the fact that you have just won 2 dollars. Putting this a little differently, your total expected earnings for the evening is no longer 0, but is 2 dollars! This can of course be generalized to the statement that your expected earnings for the evening is exactly your earnings at that point in the evening (I am of course including losses here as negative earnings). When you quit for the evening, your expected earnings of course is then your actual earnings.

Table 8.1 shows this situation for a 4-coin-flip evening. There are 16 possible scenarios. Of these 16, only 4 show heads for the first two games. This agrees with the simple calculation that getting 2 heads from 2 coin flips has a

TABLE 8.1. Expected Winnings when First Two Flips Are Won

	Coin		
1	2	3	4
H	H	H	H
H	H	H	T
H	H	T	H
H	H	T	T
	...		
H	T	H	H
H	T	H	T
H	T	T	H
H	T	T	T
T	H	H	H
T	H	H	T
T	H	T	H
T	H	T	T
T	T	H	H
T	T	H	T
T	T	T	H
T	T	T	T

1/4 probability. These 4 games are shown on top of the table, and the remaining 12 games follow. Once the first two games are won, these latter 12 games are irrelevant and we will only consider the first 4 games. Now look at the last 2 flips of these first 4 games. They show winnings of +2, 0, 0, −2, leading to the "expected expected value" of 0. The evenings game winning possibilities are then +2, 2, 2, 0, with an expected value of +2.

Looking at this situation by eliminating lines from a table leads to a proper interpretation of the situation. Winning the first 2 (or any number of) games does not change the future possibilities. I'll say it again: Coins have no memory or knowledge of each other's history. What winning the first 2 games does is eliminate from the entire list all of the possible games in which you did *not* win the first two games.

Let's go back and look at your 95% confidence interval, assuming you just won 5 dollars. It's the same curve as before except that 5 dollars has been added to every point. n now means the number of coin flips ahead of you, not counting the 5 flips that have already won you $5. Figure 8.2 shows only the part of the curve for small values of n.

For $n = 5$, we have the case that the 95% confidence interval is from (approximately) 0 to +10. In other words, if you have already flipped the coin 5 times and have won 5 dollars, and then plan to flip the coin 5 times more, you can be very confident (in this case absolutely confident) that you will go home with somewhere between 0 and 10 dollars profit. This is a quantified

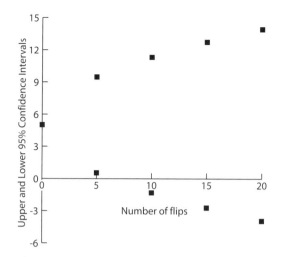

Figure 8.2. Ninety-five percent confidence interval of winnings versus number of coin flips after winning $5.

statement of common sense: If you have $5 in your pocket that you didn't start out with and you are willing to flip the coin only five times more, then there is no way that you can lose money.

Looking at this slightly differently, suppose there is a large number of people flipping coins. Let's say there are 3000 people flipping coins. The probability of someone getting 10 heads in a row is 1/1024. Using the binomial PDF formula for $n = 3000$ and $p = 1/1024$, we find that the probability of at least one person getting 10 heads in a row is 95%. This is of course what happens when you give a very unlikely event enough opportunities to occur.

The mean ($\mu = np$ for a binomial distribution) of this distribution is for 3 people to flip 10 heads in a row, with a sigma of 1.7. (The distribution is an excellent approximation of the Poisson distribution.) We can be very sure that we won't see 10 people out of the 3000 doing this (this probability is about .0007), so why not offer $200 back on a $1 bet to anyone willing to try? If we're running a big casino with many more than 3000 people passing through every day, we can be very sure of a reliable, comfortable stream of income.

The above example typifies the Las Vegas casino strategy: Set the odds (and watch the sigmas) so that you're very sure that you are both returning terrific winnings to a few people while maintaining a healthy profit margin at the same time. The fact that it's inevitable for a few people to "win big" every night seems to make a lot of people think that it's worth taking the risk to lay down their money. There's even a strong temptation to carefully watch last night's big winners and to try and understand "their system." The casinos often have loud bells and flashing lights that go off when somebody wins against these odds—just to put the idea into everybody's head that they could be the next big winner.

Now let me look in a different direction. Suppose I have a *funny* coin that has a probability of 60% of landing heads-up. Playing the same game as above, the expected winnings are just the expected value per game times n, the number of games:

$$\text{winnings} = n((.6)(\$1) + (.4)(-\$1)) = n(\$0.20)$$

This looks pretty good. The more you play, the more you win. Figure 8.3 is a graph of the upper and lower 95% confidence intervals (again, I show solid lines rather than points because there are so many points).

This is a gambler's dream: Except for the cases of just flipping less than about 100 times, you can be very confident that you won't lose money. For that matter, the more times you flip this coin, the better off you are.

Continuing on with the funny coin, let's play this same game with one additional constraint. You live in the real world, that is, you don't have infinite resources. What if you showed up at the casino with very little money in your pocket (a small "bankroll")? You know that if you can stay in the game long enough, the statistics make winning just about inevitable. But, what is your risk of getting "wiped out" before you have won enough to not have to worry?

Figure 8.4 shows your probabilities in this situation, assuming that you start out with only $1 in your pocket. N is the number of times you flip the coin before you "call it a night." For $n = 1$ there's a 0.4 probability of getting wiped out. This makes sense: If the probability of winning with this coin is 0.6, then the probability of losing 1 flip has to be $1 - 0.6 = 0.4$ For $n = 2$ there's also a 0.4 probability of getting wiped out. This is because with 2 coin flips there are 4 possibilities, 2 of which wipe you out after the first flip—and the probability of a first flip wiping you out is 0.4.

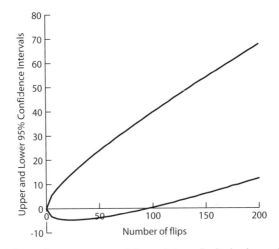

Figure 8.3. Ninety-five percent confidence interval of winnings when $P = 0.6$.

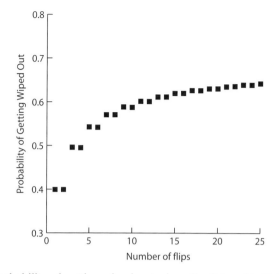

Figure 8.4. Probability of getting wiped out when $P = 0.6$ and initial bankroll = 0.

TABLE 8.2. Probability of Getting Wiped Out in a $P = 0.6$ Game After a Large Number of Flips

Bankroll ($)	Wipeout Probability
1	.43
2	.28
3	.19
4	.12
5	.09
...	
10	.01

As n increases, the probability of getting wiped out also increases. The pattern of the probability for any odd number being the same as the probability for the next even number continues. The overall probability of getting wiped out levels off at about 0.66. This leveling out occurs because the expected value of your winnings in this game increases with increasing n, so it is probable that your bankroll is growing and after a while you develop an *immunity*—if you haven't been wiped out already, you probably won't be.

The graph of your probability of getting wiped out versus the number of coin flips, looks the same regardless of your starting bankroll; but the leveling out, or *asymptotic* value of the probability for a large number of flips drops rapidly (with the number of coin flips). Table 8.2 shows this plateau value for difference amounts of initial bankroll.

The overall message thus far is pretty clear: If the odds are in your favor, you can expect to win in the long run. However, you need "deep enough pockets" to cover short-term uncertainties. This message has a parallel in business start-ups: You could have a great idea and have put together a great functioning operation (analogous to saying that the odds are in your favor), but there are almost always uncertainties that can be pretty devastating when you're just starting to build up a customer base and ship product (or whatever it is you're doing). If you aren't well enough capitalized to carry you through the rough days until you've made enough money to carry you through these rough days, you could very well get wiped out. Lack of adequate capitalization is often cited as one of the causes, if not the prime cause, for early failure of new business ventures. Unfortunately, people get swept away with the clarity of their plans and market understanding, and they often don't make adequate accommodations for the fact that so many variables have statistical fluctuations (aka a nonzero sigma) attached to them.

From the gambling casino's point of view, *they* are the people with the odds in their favor, and they have very, very big bankrolls (aka deep pockets) to carry them through statistical fluctuations. This is a very good business to be in.

THE GANTT CHART

Another example of this same issue in business planning is the Gantt chart. The Gantt chart is a chart with time as the horizontal axis and various activities needed to achieve some business goal (e.g., bringing a new product to market or getting a new factory up and running). A very simple Gantt chart is shown in Figure 8.5.

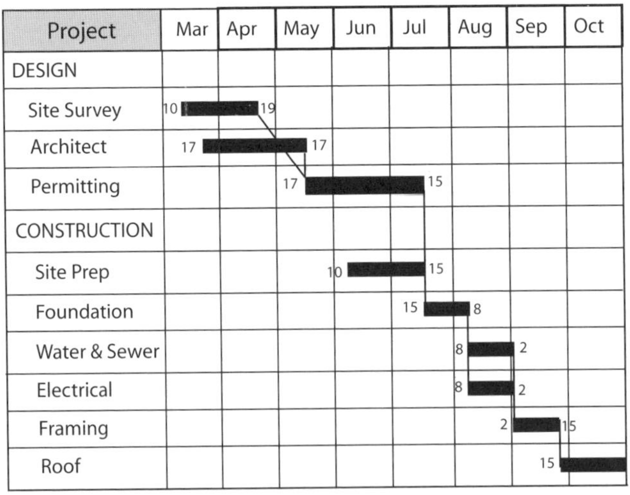

Project	Mar	Apr	May	Jun	Jul	Aug	Sep	Oct
DESIGN								
Site Survey	10 ▬▬ 19							
Architect	17 ▬▬	17						
Permitting		17 ▬▬	15					
CONSTRUCTION								
Site Prep			10 ▬▬	15				
Foundation				15 ▬ 8				
Water & Sewer					8 ▬ 2			
Electrical					8 ▬ 2			
Framing					2 ▬ 15			
Roof						15 ▬		

Figure 8.5. A simple Gantt chart.

For complicated projects such as getting a high-tech factory up and running, the Gantt chart is often an assemblage of dozens of subcharts and the total package can be a very formidable document.

The Gantt chart is a terrific management tool. Across the top is a dateline. Down the left side is a list of tasks. The bars show the planned start and completion dates for each task. The numbers on either side of the bars above tell the actual dates of the beginning and end of each task. The lines connecting the ends of tasks to the beginnings of other tasks are dependency lines—that is, indicators of what activities can't start until previous activities have finished.

In this very simple example, the site survey, the architect, and the site prep could start at any desired date. The dates shown in this example are of course arbitrary. Permitting, on the other hand, cannot begin until the site survey and the architect have completed their tasks.

Realizing that in business, time is indeed money, the Gantt charts can be very useful in predicting when a project will be completed and when different resources (architects, site excavation crews, etc.) will be needed.

But wait a minute: This chart was drawn before the project was started. Do we know, for example, that the survey (the top item on the chart) will be completed on April 19th? What if it rains suddenly and the schedule slips? What if the surveyors have a car accident on their way to work one day? What if the surveyors' previous job runs late and this job doesn't start on time? The best we can do is predict (from experience most likely) that the survey activity will take 40 days with a standard deviation of, say, 4 days and that the survey activity will begin on March 10 with a standard deviation of, say, 2 days. In this sense, Gantt chart planning can be looked at as related to gambling. The "prize" here isn't winning a certain amount of money, it's completing the job on (or before) a certain date.

Putting a sigma on every activity doesn't change the predicted completion date. It does, however, give us a sigma for how well we know this date. This, in turn, lets us calculate our confidence interval for the date of completion of the job. The answer may be painful to senior management and they will stamp their feet, throw coffee cups, and threaten to "bring in someone else who can tell us that this job will be done on time." In actuality, unless the information going into the calculations are changed, all that the new person will be able to do is to "tell us" that the job will be done on time, he or she won't be able to change the reality at all.

Here, unfortunately, is where things all too often really get funny in the upper-management offices. Since the final distribution is probably more-or-less normal, if the input information is good, then there's a 50% probability that the job will be done early rather than late. There's a 25% probability that a manager will be able to bring in two projects in a row that are early. This manager will look better than two managers who each brought in one project early and one project late, and he will look a whole lot better than a manager who brought in two projects in a row late (note of course that all four cases are equally likely). This first manager will certainly be promoted and given more responsibility based on his "track record."

Isn't chance, aka dumb luck, wonderful? A more modest conclusion would be that pure elements of chance can often weigh seriously into the results of what at first glance appear to planned, deterministic events, and we shouldn't be too eager to assign blame or reward for some of the variations.

Returning to our coin flip games, what about looking at the issues of starting with finite resources using our honest, fair, 50% heads—50% tails coin? But wait a minute, haven't I already done that? Mathematically, there's no difference between looking at what's ahead when you've won your first 10 flips and looking ahead when you start out with $10 in your pocket. Mathematicians refer to this as "reduced to the problem previously solved," and they move on to the next problem.

THE "ULTIMATE WINNING STRATEGY"

My next example is a game that you seemingly can't lose, but for some reason you just can't seem to win. This game was introduced briefly in the Introduction.

Using your fair coin, bet $1 on heads. If you win, you've won $1. Call this a success and say that the game is over. If you lose, bet $2 on heads and try again. If you win, you've covered your earlier loss of $1 and have won $1. Again, call this a success and say that the game is over. If you lose this second flip, bet $4 on heads and try again. If you win, you've covered the $1 loss of the first flip, covered the $2 loss of the second flip, and have won $1. Once again, call this a success and say that the game is over. Keep this procedure running until you've won your $1.

This scheme is unbeatable! How long can it take before you get a head when flipping a fair coin? In a short time you should be able to win $1. Then start the whole game over again, looking only to win (another) $1. Keep this up until you're the richest person on the earth.

Now let's bring in the heavy hand of reality whose name is *you can only start with finite resources.* Let's say that you walk into a game with $1 in your pocket. If you lose, you're broke and have to go home. You can't flip the coin a second time. In order to be able to flip the coin a second time, you need to have $1 for the first flip and $2 for the second flip, or $3. If you lose twice, you need an additional $4 for the third flip; that is, you need to have $7. For a fourth flip (if necessary) you need to have $15, and so on.

The need to walk into the game with all this money to cover intermediate losses is alleviated a bit if you win a few games at the outset—if you start out with $1 and win 2 games immediately, you now have $3 in your pocket and can withstand a game that requires 2 flips before winning, and so on.

Putting all of this together, we get an interesting set of curves. Figure 8.6 shows the probability of walking away with up to $10 if you walked into the

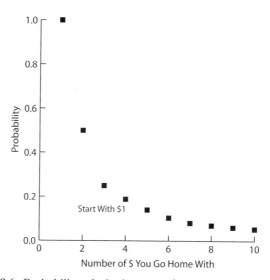

Figure 8.6. Probability of winning up to $10 when starting with $1.

casino with $1 in your pocket. Another way of putting this is the probability of winning up to 10 of these games in a row if you walked into the casino with $1 in your pocket.

The first point on the left of Figure 8.6 is the probability of walking away with $1. This is pretty much a no-brainer. You simply walk in to the casino, turn around, and walk out. There's no way that you can lose your money, so your probability of success is exactly 1.

The second point is the probability of walking away with $2. You flip a coin once. If you lose, you're broke and you must go home. If you win, you have $2 and you choose to go home. Your probability of success here is exactly 0.5.

The remaining points are the probabilities of walking away with $3 or more. As can be seen, these probabilities decrease significantly as the amount of money you hope to win increases. Note that these curves aren't obvious: If you've won, say, 5 games, then you have $5 more in your pocket than you started with, so it's harder to wipe you out, and this must be taken into account when calculating the probability of winning 6 or more games. For example, suppose you want to go home with $3. Remember that you're betting $1 every time. If you win twice in a row, you have $3. If you win, then lose, then win twice you have $3. For that matter if you win, then lose, then win, then lose, then win, then lose, ..., then win twice, you have $3. As you can see, the calculation gets quite involved. There are an infinite number of ways to win $3; fortunately, the more circuitous the path, the less likely this outcome. I chose to use a Monte Carlo simulation to generate the data.

Now I'll repeat the last graph, but add a second set of points the probability of winning up to $10 if you start out with $3 instead of $1 (Figure 8.7):

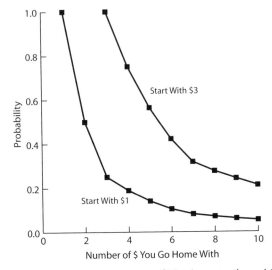

Figure 8.7. Probability of winning up to $10 when starting with $1 or $3.

The first case of the new set of points mimics the same case for the original set of points, if you want to leave with what you walked in with, just turn around and leave (100% certainty). If you want to just add $1 to your pocket, flip the coin just once or twice and you'll have a 75% probability of getting your wish. This trend continues for the whole curve. This is because you have more resilience starting out with $3 than you did starting out with $1 and you can expect better "luck". As the winnings numbers get high, however, even though you are better off starting with $3 than you were starting with $1, the probabilities still get very small.

Figure 8.8 shows what to expect if you walk in with $10 in your pocket (remember—you're still starting out with a $1 bet).

Looking at Figure 8.8, you can see that the probability of going home with $11 is very high. This is because you came in with $10 and therefore had a lot of money to back a very modest goal. On the other hand, the probability of doubling your money (walking away with $20) is less than 0.4.

Let's revise our way of thinking and say that we want to double our money. The question now is, Just how many dollars we should come in with and how should we proceed? We'll free up the $1 bet restriction, noting that everything scales with the initial bet. In other words, the curve for starting with $3 in your pocket and betting $1 on each coin flip is the same curve for starting with $30 in your pocket and betting $10 on each coin flip.

Figure 8.9 is a graph of the probability of doubling your starting position versus the how many dollars you start with, assuming a $1 bet. The probabilities jump around a bit because adding a dollar to your starting fund doesn't always get you to play another game; in this example it just raises expectations

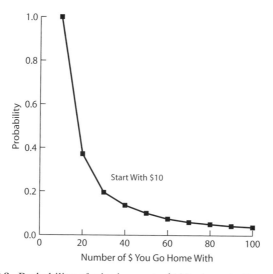

Figure 8.8. Probability of winning up to $100 when starting with $10.

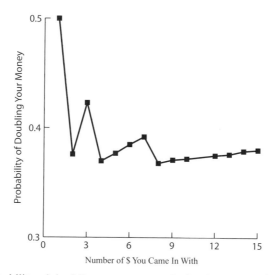

Figure 8.9. Probability of doubling your money playing bets to win $1 versus number of dollars you came in with.

(because twice your new starting amount has grown). The probabilities ultimately level out at about 0.38.

The highest probability in Figure 8.9 is 0.5, and this value occurs for the case of coming in with $1 in your pocket. Looking back at the $1 starting level curve, you see that this highest probability of 0.5 occurs for the one coin-flip situation, which is another way of saying "bet everything you have on one coin flip." It seems that after examining all the nuances of this sophisticated betting strategy, your best bet is to take all of your money, put it on the table, and flip a coin once. This whole "system" doesn't improve your odds at all!

A variation on the above game is to set the same goals, that is I want to win n_{win} dollars when I start out with n_{start} dollars. I'll play the same game (the simple coin flip) as before. However, this time I'll only bet $1 each time. I'll keep playing until either I reach my goal or I get wiped out and have to quit.

This game can be solved analytically, but it takes (or at least it took me) a few minutes of staring at some graphs to get the picture straight in my head. After each coin flip I have either $1 more or $1 less than I had before that coin flip. This is shown in Figure 8.10a for one coin flip. I start with $1. The branch going down and to the left shows me winning, I end up with $2. The branch going down and to the right shows me losing, I end up with $0. Each branch is equally likely, so the probably of traveling down any branch is 0.5 If my goal is to go home with $2 then it's clear that I have a 50% chance of meeting my goal and a 50% chance of getting wiped out. In both cases I quit for the evening after one coin flip.

Now suppose that I want to go home with $3 (starting with $1). Look at Figure 8.10b, starting at the top. If I lose the first coin flip, then I go down the right-hand branch and I am wiped out. If I win the first coin flip, then I go down the left-hand branch to the $2 point. Now I flip again. If I win the second coin flip, then I have my $3 and I quit. Since I traversed two branches, my probability of getting to this point is $(1/2)(1/2) = 1/4$.

If I lose the second coin flip, then I go down the right hand branch from the $2 point and I have $1. This is where it gets tricky—I am now at the same

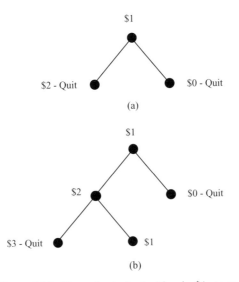

Figure 8.10. Flow charts for betting in $1 steps.

point, financially, as when I just started out and the probability of me having gotten to this point is 1/4. I can therefore write my probability of winning as

$$P = \frac{1}{4} + \frac{1}{4}\left[\frac{1}{4} + \frac{1}{4}\left(\frac{1}{4} + \frac{1}{4}\{\cdots\}\right)\right] = \frac{1}{4} + \left(\frac{1}{4}\right)\left(\frac{1}{4}\right) + \left(\frac{1}{4}\right)\left(\frac{1}{4}\right)\left(\frac{1}{4}\right) + \cdots = \frac{1}{4} + \frac{1}{16} + \frac{1}{64} + \cdots$$

This can be cleaned up using a standard trick for handling infinite series of this sort. If I multiply the above equation on both sides by 4, I get

$$4P = \frac{4}{4} + \frac{4}{16} + \frac{4}{64} + \cdots = 1 + \frac{1}{4} + \frac{1}{16} + \cdots$$

Since both of these series go on forever (are infinite), the right-hand side of the second equation is the same as 1 + the right-hand side of the first equation. Therefore, if I subtract the second equation from the first equation, I get

$$4P - P = 3P = \left(1 + \frac{1}{4} + \frac{1}{16} + \cdots\right) - \left(\frac{1}{4} + \frac{1}{16} + \cdots\right) = 1$$

If $3P = 1$, then $P = 1/3$.

This process can be repeated for any starting amount of money and any desired final goal amount. The graphs and the resulting equations get messier because there are multiple paths that lead to the starting amount, leading to multiple infinite series that must be dealt with. The results, however, are simple to express. If you start with $start and your goal is to reach $goal, then your probability is simply $start/$goal.

Just in case you lost sight of the forest for the trees, let me summarize what just happened with an example. If you start with $1 and bet $1 per coin flip and you want to end up with $3, the incorrect way of reasoning is to say that you have to win two coin flips so your probability of success is 1/4. This is because once you've won your first coin flip, you have an extra dollar in your pocket and this gives you some resiliency—you can withstand a loss without getting wiped out. This extra resiliency increases your probability of reaching your goal from 1/4 to 1/3.

The system of course cannot be finessed by clever strategies. For example, suppose I start out with $2 and have a goal of coming home with $10. According to the formula above, my probability is 2/10 = 0.20. If I decide to play the first night with a more modest goal of coming home with only $5, then my probability is 2/5 = 0.40. Then, taking my $5 the second night with the goal of reaching $10, my probability the second night is 5/10 = 0.50. Since the two nights' activities are independent, my probability over two nights of starting with $2 and ending up with $10 is therefore (0.40)(0.50) = 0.20, the same as if I'd tried to do it all in one night.

The messages of all of this chapter so far are easy to summarize:

In a fair game (expected value = 0), you're playing with equal odds of winning and losing—there is no system or strategy that can help you.

In a game with an expected value greater than 0 you will ultimately win, but make sure you are starting with enough in your wallet to cover variations as you get started.

If you are lucky and win the first few times you play, then your prospects for the rest of the evening look better than they did when you first walked in.

The expected value is the most likely result of any game (and here we're considering the Gantt chart as a type of game), but the confidence interval tells you just how sure you really are of realizing this value.

THE GAME SHOW

Another category of gambling is game-show contestant gambling. In a game show the contestant isn't risking losing any money, but rather is trying to optimize the amount of money won. Games requiring a particular knowledge (e.g., Jeopardy) are not games of chance in the usual sense. There are, however, games such as *Deal or No Deal* that are really elaborate games of chance and merit some discussion of a possible strategy. It is of course impossible to discuss every possible game, so I'll use the *Deal or No Deal* game as an example. At the time of this writing, the show is on national TV weekly. There are internet sites that describe the game and offer the ability to play the game over the internet—just for fun. There also seem to be gambling sites with a version of the game for real money, but I decided to stay away from these sites.

Here is how the game works. There are 26 suitcases (or envelopes, or boxes, etc.), each containing a different one of the 26 dollar amounts shown in the left hand column of Table 8.3. The amount in each suitcase is a secret, the suitcase is only identified by a number from 1 to 26. I have put the amounts in ascending order in the table just for convenience, you (the contestant) have no idea which amount is in which suitcase.

The game starts with you choosing one of the suitcases as your "own." This suitcase is set aside without being opened. For this example, assume you chose suitcase #20 (the one with $50,000 in it). You then choose 6 of the remaining (25) suitcases which are then opened and their dollar value revealed. These six amounts are removed from the list. In column 2 of Table 8.3 I show six amounts removed. The choice of these amounts, and all of the choices and results to follow, come from my actually playing this game on the internet several times.

What you now have is 26 − 6 = 20 suitcases unopened, one of which is set aside as "yours." At this point a mysterious character called *The Banker*

TABLE 8.3. Sample Deal or No Deal Game

Starting $	Round: 1 Cut 6	2 Cut 5	3 Cut 4	4 Cut 3	5 Cut 2	6 Cut 1	7 Cut 1	8 Cut 1	9 Cut 1
$0.01									
$1	$1								
$5									
$10	$10	$10							
$25	$25	$25	$25	$25					
$50	$50	$50	$50	$50	$50	$50			
$75	$75	$75	$75	$75	$75				
$100	$100	$100							
$200									
$300	$300	$300	$300						
$400	$400	$400	$400						
$500	$500	$500	$500	$500	$500	$500	$500		
$750									
$1,000	$1,000	$1,000	$1,000	$1,000	$1,000	$1,000	$1,000	$1,000	$1,000
$5,000									
$10,000	$10,000	$10,000							
$25,000	$25,000	$25,000	$25,000						
$50,000	$50,000	$50,000	$50,000	$50,000	$50,000	$50,000	$50,000	$50,000	$50,000
$75,000	$75,000								
$100,000	$100,000								
$200,000	$200,000	$200,000							
$300,000	$300,000								
$400,000									
$500,000	$500,000	$500,000	$500,000	$500,000	$500,000	$500,000	$500,000	$500,000	
$750,000	$750,000								
$1M	$1M	$1M	$1M	$1M					
Your Case									

makes you an offer. You can accept the amount of money you are offered and go home or reject the amount and continue to play the game.

If you reject the offer, this entire procedure is repeated a total of 9 rounds. Going forward, assuming you keep rejecting the offers, at succeeding round you choose to open (and eliminate) 5, 4, 3, 2, 1, 1, 1, and 1 suitcase up through 9 rounds. There are now 2 unopened suitcases left—yours and the one remaining from the original group of 25. If you reject the ninth offer, you get to keep what is in your suitcase.

Table 8.4 shows a full 9-round game assuming that you reject every offer. Case number 20 was your case, so you go home with $50,000. Remember that the table shows your case worth $50,000 quite clearly, but you didn't know this until after round 9.

Table 8.4 is a repeat of Table 8.3 with the following rows added to the bottom to help me analyze the game:

Nr is the number of suitcases remaining at the end of each round.

Exp is the expected value of the remaining suitcases—this will be discussed below.

Offer is the amount the *Banker* offered at the end of each round.

Ratio is the ratio of the *Banker's* offer to the expected value.

At the beginning of the game or after any round of the game, all of the choices (not yet eliminated) are equally likely. This means that the expected value of the result is just the average of the values of all of the choices. Looking at the bottom of the first column of Table 8.4, the expected value at the outset of the game with all 26 suitcases still in contention (to the nearest dollar) is $131,478.

Playing round 1, I eliminated 6 cases. On the TV show the contestants stare wisely at 25 lovely models, each holding a suitcase, and pick 6 using some algorithm that escapes me. Since nobody knows which suitcase contains what amount of money, they might as well just start at one end and work their way to the other end.

In any case, after round 1 there were 20 suitcases left, with an expected value of $119,164. The Banker offered to buy me out for $32,351. This is only about 22% of the expected value of the remaining unopened suitcases! Why would I want to take an offer that low? I suspect the answer is that the Banker knows that nobody would take an offer that low. It's just poor TV to have a contestant take a buyout offer early in the game and leave, right after the Host has spent 10 minutes getting the audience to "know" the contestant and how he or she would spend the money if they won it.

Continuing through the rounds, the ratio of Offer to Expected Value slowly grows. This has the psychological effect of really making the contestant sweat a bit before making the later decisions.

After playing this game on the internet several times, I found the pattern of this ratio starting at about 20% and slowly growing toward 100% to follow,

TABLE 8.4. Sample Deal or No Deal Game with Calculations

Starting $	Your Case	Round: 1 Cut 6	2 Cut 5	3 Cut 4	4 Cut 3	5 Cut 2	6 Cut 1	7 Cut 1	8 Cut 1	9 Cut 1
$0.01										
$1		$1								
$5										
$10		$10	$10							
$25		$25	$25	$25	$25					
$50		$50	$50	$50	$50	$50	$50			
$75		$75	$75	$75	$75	$75				
$100		$100	$100							
$200										
$300		$300	$300	$300						
$400		$400	$400	$400						
$500		$500	$500	$500	$500	$500	$500	$500		
$750										
$1,000		$1,000	$1,000	$1,000	$1,000	$1,000	$1,000	$1,000	$1,000	$1,000
$5,000										
$10,000		$10,000	$10,000							
$25,000		$25,000	$25,000	$25,000						
$50,000		$50,000	$50,000	$50,000	$50,000	$50,000	$50,000	$50,000	$50,000	$50,000
$75,000		$75,000								
$100,000		$100,000								
$200,000		$200,000	$200,000							
$300,000		$300,000								
$400,000										
$500,000		$500,000	$500,000	$500,000	$500,000	$500,000	$500,000	$500,000	$500,000	
$750,000		$750,000								
$1M		$1M	$1M	$1M	$1M					
Nr	26	20	15	11	8	6	5	4	3	2
Exp	$131,478	$150,623	$119,164	$143,395	$193,956	$91,938	$110,310	$137,875	$183,667	$25,500
Offer		$33,317	$32,351	$51,006	$75,081	$41,325	$64,138	$101,145	$178,000	$25,500
Ratio		0.221	0.271	0.356	0.387	0.449	0.581	0.734	0.969	1.000

See text for explanations of Nr, Exp. Offer, and Ratio.

but with small variations that I cannot explain. Perhaps a Monte Carlo simulation trying every possible strategy, or a computer program that plays out every possibility at each decision point, would give me more insight than simply looking at expected values. I don't think so. In some of the games I played, the ratio went up to or just a little past 100% after the 8th round and then dropped down to about 90% after the 9th round. After the 8th and 9th rounds there aren't that many different things that could happen, leading me to the conclusion that the Banker just moves numbers around a bit to keep people off their toes.

Going back to the table, after the 9th round I had a choice of taking the offer and going home with $25,500 or taking my chances and going home with either $50,000 or $1,000. At this point, mathematics doesn't help me. If I'm short of money and am risk-adverse, then $25,500 is a nice prize and I should take it. If I have money in the bank and like to take risks, then try my luck on getting the $50,000.

After the 8th round, my choices were more interesting. The offer was $178,000, so close to the expected value that the Banker is effectively offering me a Fair Game at this juncture. Looking at the suitcases available, I see 2 suitcases with values less than the offer ($1000 and $5000) but only 1 suitcase with a value greater than the offer ($500,000). This means that if I rejected the 8th round offer, there was a 67% probability that the expected value would have gone up rather than down, and offers seem to be tracking expected values. Shouldn't I go with the odds and reject the offer?

Let's look at the numbers. If I played round 9 and eliminated either the $1000 or the $50,000 suitcases, then my expected value would have gone up to $275,000 or $250,500. Both of these numbers are certainly bigger than $178,000. On the other hand, if I eliminated the $500,000 suitcase (as it turns out I did), then the expected value would have fallen to $25,500. My personal feeling (based on my economic status) is that while $250,000 or $275,000 would certainly buy a whole lot, $178,000 would buy almost as much, but $25,500 wouldn't buy nearly as much and wouldn't have nearly as much of an impact on my life. I would have taken the offer after the 8th round.

In summary, playing the game totally by the mathematics makes the choices simple: So long as the expected value is higher than the offer, keep playing. If you got to play the game hundreds of times, this would certainly be the best strategy. If you get to play this game just once, then you have to factor in how you'd feel the next morning; waking up, looking at the check you brought home the night before and deciding how to enjoy your winnings.

PARIMUTUEL BETTING

In all of the gambling games discussed so far, the odds were fixed before you started playing. In the case of the roulette wheel, we can calculate the odds.

TABLE 8.5. Parimutuel Betting Example

Horse	Bets ($)	Return per $ Bet
1	2,000	50.00
2	6,000	16.67
3	8,000	12.50
4	60,000	1.67
5	22,000	4.55
6	5,000	20.00
7	6,500	15.38
8	500	200.00

In the case of a slot machine, we (usually) don't know the odds exactly, but you can be sure that the machine owner knows them.

Parimutuel Betting, however, is based on an entirely different approach to helping you part with your money. At a horse race track, for example, the odds of a given horse winning are estimated beforehand. The actual odds are calculated using a Parimutuel Machine, today just a(nother) program on a computer, once all the bets are placed. Consider the example shown in Table 8.5.

In this table I show a horse race with 8 horses entered. The second column shows the amount of money bet on each horse. The sum of all of these bets is $110,000 (I conveniently chose the numbers to add up to a nice round number for the sake of the example, but this isn't necessary or even likely). Now I'll assume that the race track owners want $10,000 income from this race to cover their costs and profits. In practice the track owners' share is probably determined by a combination of a fixed amount plus some percentage of the bets. It isn't necessary to understand how the track owners came to their number, we just need to know the number.

After deducting their $10,000 from the $110,000 betting revenue, the track owners have $100,000 left to distribute. This is very different from, say, a roulette wheel game where since the odds are fixed before the bets are placed, the gambling casino owners only know how much they will distribute (and how much they will keep) each night as a statistical generality.

Column 3 shows how much money is paid to each gambler, per dollar of bet if her horse wins. The number in column 2 times the number in column 3 (for example, $60,000 × 1.67 for horse #4) always equals $100,000. This is just another way of repeating that the full $100,000 is always paid out to the betters.

The number in column 3 is also identically the odds on this horse as a ratio to 1. For example, the odds on horse #1 was 50:1. The odds on horse number 4 is 1.67:1; this is more usually quoted as the equivalent ratio of two integers (multiply both numbers by 3) 5:3.

Now, "odds" has been defined as another way of quoting a probability with both representations being valid (sort of like quoting a cost in dollars and euros). In this situation, however, the meaning of discussing "odds" or "probability" of something after it's happened is a little hazy. Probability is the likelihood of an event chosen out of the allowed list of random variables with a given PDF occurring *in the future*. If I flip 10 coins and get 10 heads, does it really mean anything to discuss the probability of it happening? It has already happened. When we believe that probabilities don't change in time, as in the probability of a coin flip, then we can use history to teach us what the probability is going forward. We're often a little sloppy with the language and will say something like "The probability of you having flipped 10 coins and gotten 10 heads was only 1/1024." What we really mean is "The probability of you flipping those 10 coins again and getting 10 heads is 1/1024." Something that's already happened doesn't really have a probability, since it already happened, it's a certainty that it happened. (This is a funny sentence structure, but it is nonetheless accurate.)

Returning to the issue at hand, before a race there will be estimated odds published. These are best guess probabilities. After the race, the "odds" are really just a statement of how much money you would have won per dollar betted if your horse had come in.

Looking at the table again, we see that horse #4 didn't pay very much if it won. This is because so many people thought that horse #4 would win (or, equivalently, a few people were so sure it would win that they bet a lot of money on it) that the $100,000 winnings had to be spread very thinly. Horse #4 was indeed the "favorite" in this race.

On the other hand, almost nobody bet on horse #8. This horse was a "long shot"; if it won, it would have made a few betters very happy.

From the perspective of the race track owners, this is a terrific system. So long as enough people show up to cover the owners' cut plus some reasonable amounts to pay back to the betters, the track owners know exactly how much money they'll make every day. They never have to worry about a few slot machines "deciding" to pay off an awful lot of money one day.

Related in concept to parimutuel betting is the concept of *value pricing*. Value pricing is used by airlines and some hotels to adjust prices for their products as a function of the immediate market demand for the product. This is a significant extension of the idea that a resort hotel will have different "in-season" and "out-of-season" rates.

An airline will sell tickets at a fairly low price a long time before the date of the flight(s). This is because they haven't really established the need for these seats yet and also because this is one way of assuring your business and in turn giving the airline a better chance to project revenues and possibly even tweak flight schedules.

On the other hand, an airline will charge very high prices for its tickets from about 1 to 3 days before a flight. This is when the business traveler who developed a last-minute need for a trip is buying tickets; this trip can't be postponed.

In other words the demand is very high for these seats and the customers can't walk away from a high price.

Immediately before the flight an airline will release upgrades to a waiting list. For very little or no revenue the airline is promoting goodwill among its customers while invoking virtually no incremental cost—these people are already getting on the plane, so the fuel costs are constant, and the drinks and rubber chicken dinner do not really cost that much. Other price adjustments are made in response to timing and demand for seats. It's been said that no two people on a flight paid the same amount for their tickets. I doubt that this is accurate, but there certainly were a lot of different prices paid for the tickets on a given flight.

This type of pricing is related to parimutuel calculations in that the adjustments are being made in response to incoming data in real time. The computer algorithms that adjust the prices, however, are not reflecting simply mathematics such as in the horse race example, but are programmed with formulas that reflect the airlines' marketing experts' view of what people will pay for a ticket in different circumstances (and also in reaction to what the competition is charging). In other words, while there really isn't a parimutuel machine setting the prices, it just seems that way when you're trying to shop for tickets.

CHAPTER 9

TRAFFIC LIGHTS AND TRAFFIC

The decisions we make when driving down the road are incredibly sophisticated. Start by assuming that you're the only car on the road, so you don't have to worry about collisions, picking the best lane, passing or being passed, and so on. Based on your car and the posted speed limit, you have a maximum speed that you are willing to cruise at. When pulling away from a traffic light, you have a rate of acceleration that you're comfortable with as being not fast enough to snap your neck back, but not so slow that you feel that you're wasting time. Similarly, when pulling up to a (red) traffic light, you have a braking, or deceleration, rate that you're comfortable with.

As you're driving, you look up ahead and see that the next traffic light has just turned from green to yellow. How will you react? If the traffic light is very far away, you'll estimate that the light will go from yellow to red to green by the time you reach it, so there's no point in slowing down at all.

If you're very close to the light when it turned yellow, you'd probably assume that you would be past the light or at least be into the intersection before the light turns red, so again you don't slow down at all.

In between these two extremes is the situation where you think the light will have turned red by the time you reach it, so you slow down and then come to a stop at the traffic light. You wait for the light to turn green, accelerate to your cruising speed, and continue on your journey.

Probably Not: Future Prediction Using Probability and Statistical Inference,
by Lawrence N. Dworsky.
Copyright © 2008 John Wiley & Sons, Inc.

OUTSMARTING A TRAFFIC LIGHT?

Have you ever wondered if there might not be a better way?[1] Let me run through a possible scenario. You're driving down a road with, say, a 40-mile-per-hour (MPH) speed limit. Being a conscientious citizen, you never exceed the speed limit; that is, you're driving at exactly 40 MPH. When you do start or stop, except in case of emergencies, you accelerate and decelerate conservatively. I'll assume that you accelerate from 0 to 40 MPH, at a constant acceleration, in 7 seconds. You slow down at this same rate.

Along this road is a traffic light. This traffic light has a red light duration of 30 seconds. Shortly I'll be discussing a road with many lights and the overall impact of all of these lights on the progress of traffic down the road. Right now I'm just considering one light with a red duration of 30 seconds. Also, there are no other cars on the road.

Depending on just when you come along with respect to the timing of this traffic light, you might be lucky and not have to stop or even slow down at all. On the other hand, if you're unlucky, you'll have to come to a stop and wait for the light to turn green.[2]

If you have to stop at the light then you are delayed by the period of time you have to wait until the light turns green, and then slowed by the time it takes for you to accelerate up to your full speed of 40 MPH.

While accelerating at a uniform rate (of acceleration) from 0 to 40 MPH, your average speed is just 20 MPH. This may be seen by examining Figure 9.1. Since the acceleration is constant, the speed is increasing linearly (the graph of speed versus time is a straight line). Think of the line as being made up of a large number of points. This is shown in Figure 9.1 as points every half-second. For every point above the center of the line in time (the point at $t = 3.5$ seconds) there is a corresponding point below the center of the line. The difference between the speed at $t = 3.5$ seconds (20 MPH) is the same amount above 20 MPH for the first point as it is below 20 MPH for the second point. The average speed for these two points is therefore 20 MPH. This implies that the average speed for all the pairs of points must be 20 MPH and therefore the overall average speed is 20 MPH.

---

[1] A strong word of (hopefully obvious) warning is in order here. I'm presenting these simulations as an exercise in seeing how the probabilities of certain happenings and your knowledge of the properties of these probabilities might give you insight into future events and in turn show you how to optimize your decisions. In this chapter I'm idealizing many things in order to present what I hope is interesting fuel for thought. Sometimes the results of the calculations shouldn't really be put into practice. For example, it certainly is not a good idea to drive through an intersection at full speed just as the traffic light turns green—even though the mathematics tells you that this will get you home fastest without running a red light.

[2] This is a classic situation where our memories rarely accurately reflect reality. If the light is green half time and red the other half of the time, then this light should impede our progress only half the time. Most of us, however, would be willing to swear that this light is inherently evil, sees us coming, and almost always turns red just before we reach it.

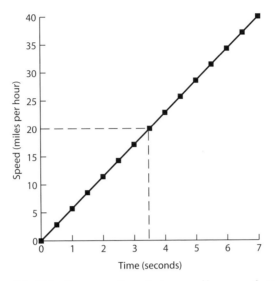

Figure 9.1. Velocity versus time during a uniform acceleration.

If for the first 7 seconds after the light turns green you are traveling at an average speed of 20 MPH, then you are only traveling the distance you would have traveled in 3.5 seconds had you been going your full speed, 40 MPH. You therefore have to drive for an extra 3.5 seconds at 40 MPH to make up this time.

Now I'd like to make an assumption of something that's unlikely but not physically impossible. I'd like to assume that you started timing the light when you saw it turn red, you know that the red light duration is exactly 30 seconds, and therefore you know exactly when the light will turn green. You stop exactly the distance before the light that you would travel in 7 seconds at an average speed of 20 MPH. Then, exactly 7 seconds before the light is to turn green, you put your foot on the gas and accelerate at your usual rate. You will reach the light just as your speed reaches 40 MPH and exactly at the time that the light turns green.

Since all the time while waiting at the red light is wasted in the sense that you cannot go past the light during this time, what you are doing here is absorbing the 3.5 seconds lost in accelerating to full speed into this wasted time. Note that the above strategy is not unique. While you do not want to accelerate faster than your usual rate (screeching tires and all that), you certainly can accelerate slower than your usual rate. Figure 9.2 shows the speed and position versus time for the conventional stop-at-the-light-and-wait-for-it-to-change strategy.[3] Figure 9.3 shows the improved strategy as described

---

[3] I haven't derived the fact that position increases with the square of time during uniform acceleration whereas it increases linearly with time when there's no acceleration. This requires a bit of integral calculus applied to Newton's $F = ma$.

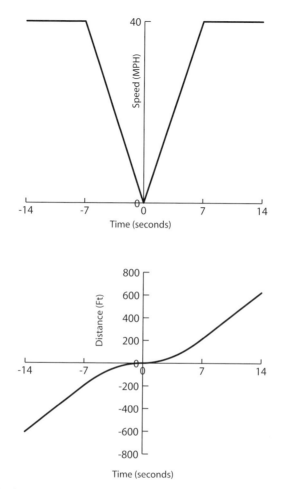

Figure 9.2. Velocity and position versus time when stopping and waiting at a red light.

and Figure 9.4 shows another strategy that is equivalent in terms of time saved. The difference between Figures 9.3 and 9.4 is that in the latter strategy the deceleration is started sooner and is more gradual. In all three of these figures the traffic light is at $x = 0$ and the light turns green at t (time) = 0. The results of this strategy show in these figures as the extra distance traveled at $t = 14$ in Figures 9.3 and 9.4 as compared to the distance traveled at the same time in Figure 9.2. The strategies shown in Figures 9.3 and 9.4 are of course not unique; any deceleration/acceleration scheme that puts you at $x = 0$ at time $t = 0$ and going full speed is equivalent in saving the maximum possible amount of time.

Is this strategy worthwhile? Probably not. Furthermore, it's not really work-able because you're rarely in the position to predict exactly when the light will

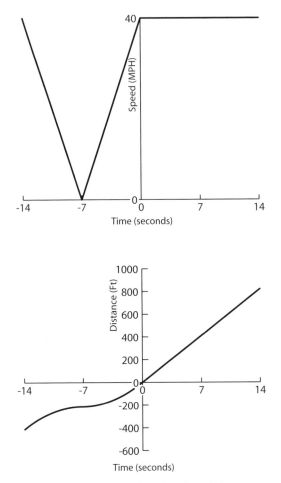

Figure 9.3. Velocity and position versus time for the minimum wasted time at a red light strategy.

turn green. If you can describe the light-switching time with a uniform distribution, then you can put together a strategy based on the latest possible time that the light could turn green. In this case when the light turns green sooner than the latest possible time you would immediately start accelerating. Since the best possible saving in this example is 3.5 seconds, it doesn't take much uncertainty in the light changing time before your expected saving is very small.

If the best you could do is to predict the light changing time with a normal distribution, then things get worthless very quickly. Even if you base your calculations on a 99% confidence factor, if you drive this route to and from work every day, 5 days a week, then about 4 times a year you would be

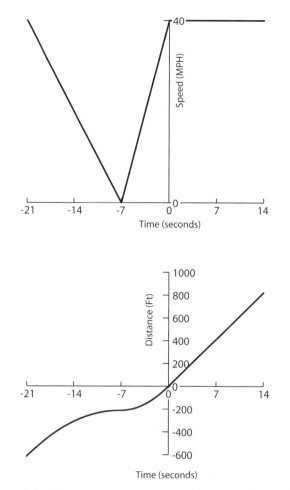

Figure 9.4. Alternate strategy equivalent to that of Figure 9.3.

accelerating into an intersection before the light has turned green. I think calling this ill-advised is an understatement.

Maybe in the future, when "smart" traffic signals communicate directly with "smart" auto control computers, it will be possible for a car to know just when a light will change and that the intersection is clear of cars being driven by someone who's a little sloppy about stopping in time when the light is just turning red. Then again, if and when such a smart system comes into universal use, it should be possible to remove the traffic lights entirely and have a control computer handling optimum vehicle control. Until that time, however, it looks like we all just have to roll up to a red traffic light and wait for it to turn green.

MANY LIGHTS AND MANY CARS

Now suppose that there's a car in front of you. I won't consider the possibility of your passing this car because there would be nothing new to discuss—once you've passed the one car in front of you, you've recreated the situation where, effectively, you're the only car on the road.

Driving with a car in front of you increases the complexity of your decision process. First of all, if the other driver's cruising speed is lower than your ideal cruising speed, then your cruising speed has just been lowered. Also, while you can pull up right behind this other car at a stop light, when cruising you have to set a safe distance between your car and the car in front of you. Furthermore, when accelerating away from a traffic light that's just turned green, most drivers wait a second or two after the car in front of them has started moving before they start moving.

Perhaps the most annoying situation when following another car is that a traffic light turns yellow as the car in front of you approaches it, the car in front of you successfully drives through the intersection and continues on its way, and you are stuck waiting at the (now) red light.

In order to study the statistics of what to expect from a street with traffic lights, first without and then with other cars, I put together a Monte Carlo simulation that puts 10 cars on a 10-mile journey. There are 20 traffic lights along the way. These lights are nominally evenly space apart, but I varied the light separation with a normal distribution about there nominal positions and a 500-foot standard deviation. The traffic lights have a 25-second green duration, a 5-second yellow duration, and a 30-second red duration, all adding up to a 60-second repeat period. The timing of when to start this period (i.e., when each light turns, say, green) is a random variable, uniformly distributed in the 60-second overall period.

For the first round of simulations, I set everybody's maximum cruising speed to be the same. Also, everybody accelerates and decelerates at the same uniform rate as the first example in this chapter, the rate which gets you from 0 to 40 MPH in 7 seconds and from 40 miles an hour back down to 0 in 7 seconds.

The road has a stop sign at the beginning, so all drivers start out at 0 MPH.

SIMULATING TRAFFIC FLOW

In this section I'll describe the algorithms that make up the traffic flow simulation. This is how a Monte Carlo simulation gets created. If you're not interested in reading this but you would like to see the results of the simulation, just skip ahead to the next section.

I'm going to assume that I have formulas for everything I need. For example, if there's a car in front of me just starting to move, I know how much space I have to give him before I start to move as well as how much separation to keep between us, depending upon how fast I'm going.

The Monte Carlo simulation consists ultimately of running the simulation many times and looking at the statistics of the results of all the runs. It is only necessary to describe one run in detail to understand the simulation. The run looks at one car at a time, starting with the first car and going down the list in order. This is because the driver of each car only bases his decisions on what's in front of him.[4] To understand the algorithms involved, it's only necessary to follow the logic for one of the cars. As you'll see below, many decisions are based upon whether or not there's a car in front of you. If you're the first car, then there is never a car in front of you and some of the paths in the flow chart simply never get taken.

I will be storing information for each car as to its position and speed at all times. I start at time $(t) = 0$, position $(x) = 0$, and speed $(v) = 0$. I chose to track time in 0.1-second steps and distance in feet. This seemed reasonable to give me the resolution I needed while not creating excessively long computer run times (these days, computer memory is inexpensive and available, so there was never an issue of not having the RAM to store all my data as I proceeded).

At each time step, the first thing I do is look at myself. Am I moving? This I find out by looking at my speed. The simple answer to this is yes or no. First I'll consider the case of "no." In this case I'd like to find out why I'm not moving. I look ahead and see what's in front of me—a car or the next traffic light.

This decision process can be followed on the flow charts shown in Figures 9.5 and 9.6. Figure 9.5 shows the initial "Am I Moving?" decision and the algorithm for "No (I am Not Moving)." Figure 9.6 shows the algorithm for "Yes (I Am Moving)." These flow charts show considerably less detail than does the flow chart in Chapter 7. My intent here was to show the logic flow that's analogous to what goes on in a driver's mind in these situations rather than to actually show all the details of the computer algorithms.

Notice that all paths on the flow chart(s) for this simulation end up in the box "Time Step." This box takes input information of the desired acceleration or deceleration. It then advances time by 0.1 seconds and, based on this time step and the desired acceleration or deceleration, calculates the new position and speed. If the new speed is faster than the allowed maximum speed, the speed gets clamped at the maximum speed. This logic handles all

---

[4]This again is a good abstraction for mathematical modeling, but not the best practice on the road. A good driver always tries to know what's going on all around him.

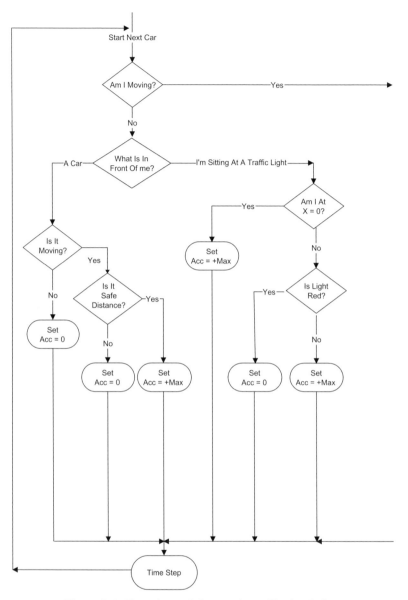

Figure 9.5. Flow chart of the cars in traffic simulation.

contingencies. For example, if you're standing still and you set acceleration to 0, this gets interpreted as "stay put for another 0.1 second." Also, slowing down (decelerating) is accomplished by stipulating the negative of an acceleration rate.

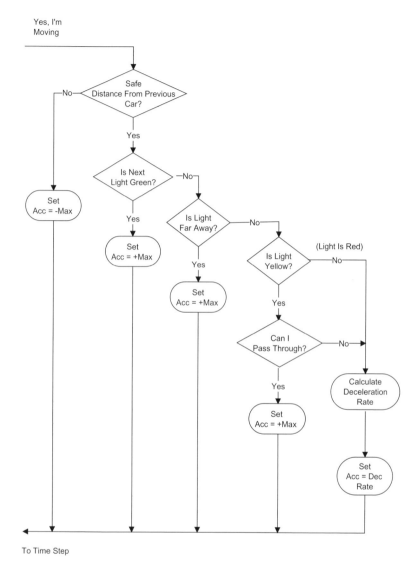

Figure 9.6. Continuation of Figure 9.5.

SIMULATION RESULTS

Figure 9.7 shows the results of one simulation run for the first car, with maximum (cruising) speed = 40 MPH. The straight line shows the position versus time for an ideal situation—a car traveling at a uniform 40 MPH through the region. At 40 MPH you'll go a mile in 90 seconds so to travel 10 miles will take 900 seconds. The actual time for this car is approximately 1100 seconds.

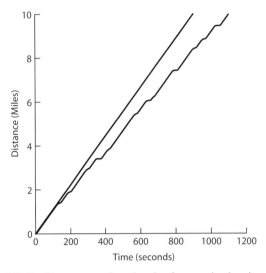

Figure 9.7. Position versus time for the first car in the simulation.

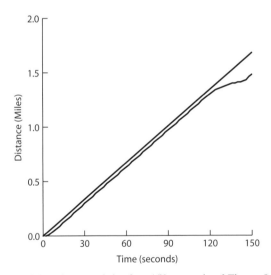

Figure 9.8. Blowup of the first 150 seconds of Figure 9.7.

Figure 9.8 is a blow-up of the first 200 seconds of Figure 9.7. Looking at this figure, we see that there is an immediate loss of distance covered (as compared to the straight line) near $t = 0$. This is because the simulation starts the car from a stop sign; that is, it has to accelerate up to full speed. Then the car travels at full speed, as can be seen by the fact that the two lines are parallel, up until $t \sim 120$ seconds. At this time the car starts slowing down for a traffic light. Since this traffic light is at ~1.4 miles into the trip and the traffic lights are about 0.5 miles apart, the first two lights must have been green. The car slows until $t \sim 150$ seconds and then accelerates to full speed again. Apparently this traffic light turned green just as (or shortly after) the car reached it. You can see real stop-and-wait situations at $t \sim 350$ and $t \sim 1000$ seconds in Figure 9.7.

The full trip took 1100 seconds. For a 10-mile trip, this means that the average speed was 32.7 MPH. On this trip with a traffic light every half-mile, the travel time increased about 22% over the light-free trip.

Now let's look at the second car. Figure 9.9 is a repeat of Figure 9.7 with the second car's information added, and Figure 9.10 is a blowup of the first 150 seconds.

Figure 9.10 doesn't show anything exciting. The second car starts moving shortly after the first car has started moving, and it follows behind the first car. If Figure 9.10 represented the whole story of what it's like to be behind another car in traffic, then there would be nothing interesting to learn. However, look what happens at $t \sim 500$ (Figure 9.9). Car 2 gets caught at a light that car 1 has already past through. From then on the two cars' progress is not as closely coupled as they had been. While they both get caught at lights at $t \sim 580$ and $t \sim 680$, at $t \sim 720$ car 2 gets caught at another light that doesn't stop car 1. On the whole, however, car 2 falls behind car 1 and never catches up. If the trip were longer, with more traffic lights, we could expect to see the distance between these cars lengthening. On this relatively short trip, the average speed for the car 2 is 31.0 MPH.

Figure 9.11 shows the simulation results for 10 cars. As may be seen, and as I'm sure we've all experienced, the more cars on the road, the longer it takes to get somewhere—even though in this situation most cars get up to full

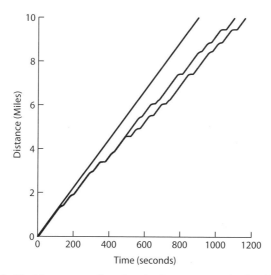

Figure 9.9. Position versus time for the first two cars in the simulation.

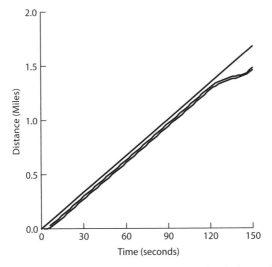

Figure 9.10. Blowup of the first 150 seconds of Figure 9.9.

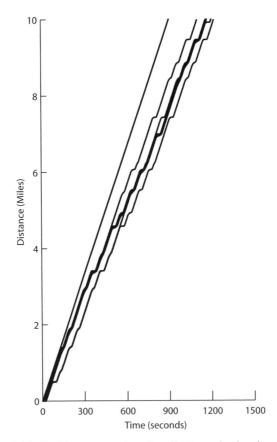

Figure 9.11. Position versus time for all 10 cars in the simulation.

speed between lights. The 10th car in this group has an average speed of 27.1 MPH. Traffic is indeed taking its toll on progress!

In Figure 9.11 it's hard to see that there are indeed 10 cars being tracked. Making the graph big enough to see everything clearly while including all of the data on one sheet would have required a fold-out sheet much larger than the dimensions of this book. If I were to blow up the region near $t = 0$, you would see the progress of 10 cars, equally spaced. On the other hand, what if I blow up a region near the end of the trip, say around 9 miles from the starting point? (Figure 9.12).

Looking at Figure 9.12, we see that car 1 is well ahead of the pack. At about 9.47 miles in his journey he stops at a traffic light. Cars 2–9 are bunched together, with car 2 coming to the same traffic light about 1 minute after car 1. Car 2 also has to stop at this light. Behind car 2, the next few cars stop at the light and the remainder of the bunch slow down because they're coming up to some cars that are/were waiting for the light to turn green and are beginning to pull away. In the figure you can see clearly the slow acceleration of

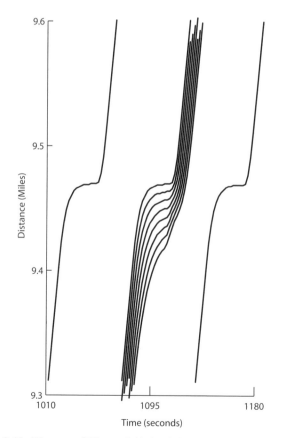

Figure 9.12. Blowup of Figure 9.11 for 9.3–9.6 miles along the road.

the later cars in the bunch as they keep safe distances between them. Finally we see car 10, about a minute behind the bunch. This car also has to stop at this light.

The bunching phenomenon is an interesting side effect of having traffic lights and is something we've all experienced. It's unfortunate in that there are probably many more accidents when several cars are traveling close to each other than there would be if they were spread apart. However, if they were spread apart, then there would be no way for traffic coming through the intersection when their light is green (i.e., when your light is red) to progress. Again, maybe a future computer controlled intelligent system will make things more efficient and safer.

So far I've presented the results of one pass of the Monte Carlo simulation. The next pass(es) should look similar but of course will not be identical to this first pass. Figure 9.13 shows the average speeds and Figure 9.14 the standard deviations for all 10 cars as a function of the maximum cruising speed. It seems that while the later you are in the queue (car 10 versus car 1), the longer you can expect the trip to take, but the less variation there is in your expected average speed. Just by looking at these figures, however, you can see that the

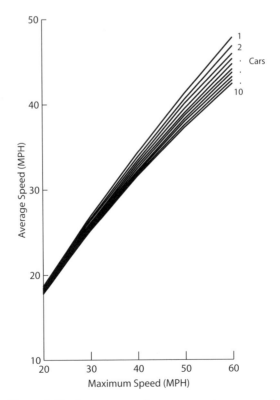

Figure 9.13. Average speeds versus maximum speed.

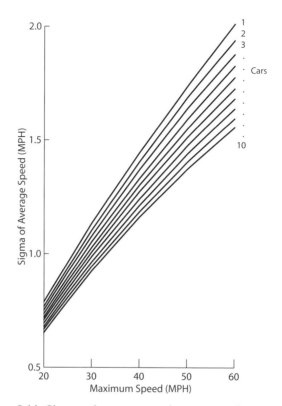

Figure 9.14. Sigmas of average speeds versus maximum speed.

standard deviations as a fraction of the average speed really don't vary that much from car to car. Figure 9.15 shows the speed distributions for a maximum speed of 40 MPH (40 MPH chosen as a representative example). Not all the cars are shown because the graph gets too cluttered to see clearly when all the cars are shown. Also, the curves are jagged because I stepped the maximum speeds in 5 MPH increments, but it is pretty clear that the distributions look normal.

Another way of looking at these data is to consider the time lost due to the traffic lights and the traffic. Our 10-mile drive at 40 MPH, for example, ideally takes 900 seconds (15 minutes). If it actually takes 1050 seconds, with the extra time due to the traffic lights and the other cars, then we can attribute 150 seconds of lost time to these traffic lights and other cars. Figure 9.16 shows this lost time for the 10 cars.

The first thing to notice in Figure 9.16 is that the lost time doesn't vary much with the maximum driving speed. For the fifth car (just as an example), the lost time varies between 178 and 205 seconds when the maximum driving speed varies between 20 and 60 MPH. The average of 178 and 205 is 191.5 seconds; 178 and 205 are just −7% and +7%, respectively, below and above

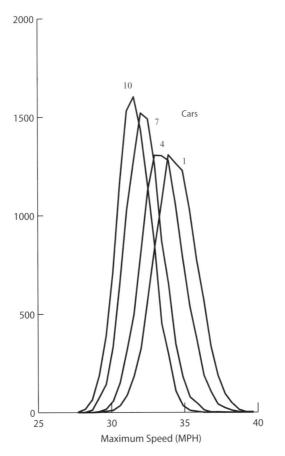

Figure 9.15. Distribution of average speeds for 40 MPH maximum speed.

the average for changes in maximum driving speed of 50% above and below the average. It seems that time lost due to traffic lights and the amount of traffic cannot be compensated for by raising the speed limit.

On the other hand, at 40 MPH (again, just a typical example), the lost time varies between 150 seconds for the first car and 239 seconds for the tenth car. Repeating the calculation of the last paragraph, these are −23% and +23% variations.

Up until this point, I've maintained a useful but inaccurate fantasy: I've assumed that all drivers will, given the chance, drive at exactly the speed limit. I say that this was a useful fantasy because it let me generate some results that are reasonable, so that we could get a picture of what's going on, without a totally nonsensical loss of accuracy.

I've never done or seen a careful study of the statistics, but my observation after over 40 years of driving is that, on roads other than highways out in the desert and the like, most people will drive at an average of about 5 MPH over

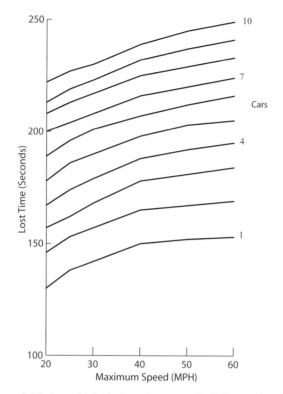

Figure 9.16. Lost (delay) time due to traffic lights and traffic.

the posted speed limit with a standard deviation of about 5 MPH about this average. I'm probably erring in making this a symmetric distribution about the average because I want the convenience of working with a normal distribution. In reality, there are probably more people driving on the high side of the average than on the low side. Nevertheless, I think that by using a normal distribution rather than a fixed maximum speed for all drivers, I'm getting much closer to reality.

Figure 9.17 shows the average speed resulting from a posted maximum speed of 40 MPH. For comparison, the same information for the previous case (everyone driving at a maximum speed = posted maximum speed) is also shown. The results are interesting in that they cross—there is no across-the-board better situation. When you're the first car in the group, then you do better if your peak speed is 5 MPH above the posted speed limit than if your peak speed is the posted speed limit. Even after averaging over drivers with all the personal variability that I discussed above, this is the case. When you're the last car in the group, on the other hand, you do much worse when all the drivers are different. This is because, statistically, there's a good chance that one of the cars in front of you (in this case there are 9 of them) will be a

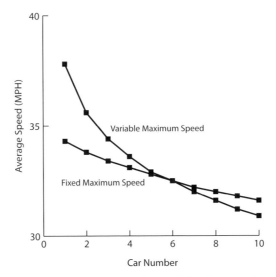

Figure 9.17. Average speeds versus posted speed for fixed and variable maximum speeds. Posted speed = 40 MPH.

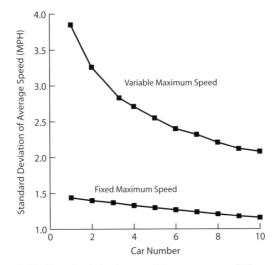

Figure 9.18. Standard deviation of the averages of Figure 9.17.

slow driver and since we're not considering passing in this simulation, you're stuck.

Figure 9.18 shows the standard deviations of the averages given in Figure 9.17. In all cases the standard deviation is much higher when each driver "does his own thing." The standard deviation falls with the number of cars, indicating that there is averaging over the different drivers taking place.

Using the awesome computer capabilities and sophisticated computer simulation programming languages available today, it is possible to accurately simulate complex traffic situations including traffic flows on cross streets, multilane roads with drivers switching lanes, passing, and so on.

This type of simulation is a valuable tool in planning roads, intersections, traffic light patterns, and so on. Also, while there are not yet on-board computers and car-to-car data exchange, there are "smart" traffic lights that sense the traffic flows and number of cars waiting to turn or go through an intersection, and so on. Every piece of information gathered "in real time" on the scene and properly reacted to can change the statistics of traffic flow tremendously—without compromising safety.

CHAPTER 10

COMBINED AND CONDITIONAL PROBABILITIES

All of the probabilities calculated up to this point have been, in a sense, *simple* probabilities. This doesn't mean that they were necessarily easy to calculate or understand, but rather that they all share some fundamental properties that will be explained below. I'd like to talk about some not-so-simple probabilities, most of which won't be that hard to calculate and understand. Once again, however, I need to introduce some notation that can capture the ideas that I'll be presenting in a concise and precise manner.

FUNCTIONAL NOTATION

In Chapter 2, I introduced the idea of a function and then went on to probability distribution functions. The common (actually, universal as far as I have ever seen) notation for a function is to use some letter followed by something in parentheses. Since I'm dealing with probabilities, I'll stick with convention and use the letter P. If I'm talking about rolling a pair of dice, I would say that the probability of the sum of the two dice equaling 12 is

$$P(12) = \frac{1}{6}$$

Probably Not: Future Prediction Using Probability and Statistical Inference, by Lawrence N. Dworsky.
Copyright © 2008 John Wiley & Sons, Inc.

This notation is a little bit confusing in that if I had some variable P that I wanted to multiply by 12, I could write it the same way. The result of course probably wouldn't be 1/6, depending on the value of the variable, P. How do you know whether I'm talking about a function or a multiplication? Unfortunately, only from context—this is one of the rare situations where the accepted mathematical notation for an operation isn't the clearest way of expressing something.

In this *functional notation*, if I defined the variable x as representing the sum of the two dice, then I could make the above somewhat clearer by writing

$$P(x = 12) = \frac{1}{6}$$

The power of the notation is the ease in which I can express more complex expressions. For example, the probability of x being 10 or more is

$$P(x \geq 10) = \frac{3}{36} + \frac{2}{36} + \frac{1}{36} = \frac{6}{36} = \frac{1}{6}$$

Let me redefine my variables a bit. I'll label the two dice separately. If I paint one die red and call its value x and paint the second die green and call its value y, then the probability of the red die equaling 1 is written as

$$P(x = 1) = \frac{1}{6}$$

In this case we are dealing with 2 dice, but since I didn't specify what the green die is, then I don't have to mention y anywhere and the above is correct.

With this notation as a working tool, I am now ready to start talking about *combined probabilities*. The first combined probability uses the logical operation *and*. I'll keep using the definitions of x and y as the values of the two dice. Then the probability that the red die will be greater than 4 and the green die will equal 2 is written as

$$P(x > 4 \text{ and } y = 2)$$

Now that I have a way to write it, I should also talk about how to calculate it. As in all of the coin flipping and dice rolling examples thus far, the two events ($x > 4, y = 2$) are independent. When events are independent, the probability of both of them happening (the *and* case) is just the product of the probabilities of the two occurrences:

$$P(x>4 \text{ and } y=2) = P(x>4)P(y=2) = \left(\frac{2}{6}\right)\left(\frac{1}{6}\right) = \left(\frac{2}{36}\right) = \left(\frac{1}{18}\right)$$

In this example you see the functional notation mixed together with multi-plication notation. It's a little confusing, but hopefully not too bad.

The next logical operator is *not*. This is a convenient way to express things in many cases and mathematically is quite simple. Since an event must either occur or not occur, but not both, then

$$P(E) = 1 - P(not\ E)$$

where *E* is an event such as a die having a value greater than 3.

The last logical operator I'll consider I actually already used, just above. This is the operator *or*. The definition of the occurrence of event A *or* event B occurring means that either event A will occur, event B will occur, or they both will occur.[1] As an example, what is the probability that die *x* is 4 or die *y* is 6?

Table 10.1 shows all 36 possible events. There are 6 cases where *x* = 4 and 6 cases when *y* = 4. If you therefore conclude that the answer is $6+6/36 = 12/36 = 1/3$, you will see that this conclusion does not agree with the last column's result of 11/36. This is because you are double counting the case when *x* = 4 and *y* = 6.

The correction is simple, don't double count. In other words,

$$P(A \text{ or } B) = P(A) + P(B) - P(A \text{ and } B)$$

There are some situations where event A and event B are *mutually exclusive*. This is a fancy way of saying that the two events simply cannot occur at the same time. Consider the example shown in Table 10.2. We want the probability that the sum of two dice is less than 4 (event A) or the sum of two dice is greater than 10. There are 3 ways to get event A and 3 ways to get event B, but no ways to get them both at once. The simple answer in this situation is therefore

$$P(A \text{ or } B) = P(A) + P(B) = \frac{3}{36} + \frac{3}{36} = \frac{6}{36} = \frac{1}{36}$$

because $P(A \text{ and } B) = 0$.

---

[1] There is also the *exclusive*-or, which is defined as either event A occurs or event B occurs, but *not* both. This is the less common usage of the term *or*, and you can expect it not to be this case unless you're specifically told otherwise.

TABLE 10.1. All the Ways to Get $P(x > 4 \text{ or } y = 6)$ from a Roll of Two Dice

x	y	$x = 4$	$y = 6$	OR
1	1			
1	2			
1	3			
1	4			
1	5			
1	6		+	+
2	1			
2	2			
2	3			
2	4			
2	5			
2	6		+	+
3	1			
3	2			
3	3			
3	4			
3	5			
3	6		+	+
4	1	+		+
4	2	+		+
4	3	+		+
4	4	+		+
4	5	+		+
4	6	+	+	+
5	1			
5	2			
5	3			
5	4			
5	5			
5	6		+	+
6	1			
6	2			
6	3			
6	4			
6	5			
6	6		+	+

TABLE 10.2. All the Ways to Get $P(x + y < 4$ *or* $x + y > 10$) from a Roll of Two Dice

x	y	A $x + y < 4$	B:: $x + y > 10$	OR
1	1	+		+
1	2	+		+
1	3			
1	4			
1	5			
1	6			
2	1	+		+
2	2			
2	3			
2	4			
2	5			
2	6			
3	1			
3	2			
3	3			
3	4			
3	5			
3	6			
4	1			
4	2			
4	3			
4	4			
4	5			
4	6			
5	1			
5	2			
5	3			
5	4			
5	5			
5	6			+
6	1			
6	2			
6	3			
6	4			
6	5			+
6	6			+

CONDITIONAL PROBABILITY

A few chapters ago I talked about the probability of winning money in a coin flip game once you had already won some money. I didn't say it at the time, but this is a case of *conditional probability*. Conditional probability considers a complex situation and looks at the probability that some event B will happen once some interrelated event A has already happened. The notation for this is

$$P(B|A) =$$

which is read "the probability of B given A," which in itself is a shorthand for "the probability of B given that A has already happened." This is significantly different from the situations where I kept emphasizing that the coin or the die has no knowledge of what the other coins or dice are doing, or even what it did itself last time. Now I'm going to look at situations where interactions and history change things. The first few examples will be simple, just to show how things work. Then I'll get into some examples of situations and decisions that could literally mean life or death to you some time in the future.

Suppose I have 3 coins to flip and I want to know the probability of getting 3 heads. Before I flip any of these coins, I can calculate this probability as 1/8. Now I flip the first coin and get a head. What's the probability that I can flip the other two coins and get two more heads? My original sample space consists of 8 possibilities (Table 10.3), and the first possibility is the only one that shows 3 heads. Now I flip coin A and get a head. This immediately eliminates the last 4 possibilities from the list and leaves the first 4 possibilities as my *reduced sample space*. There is still only one way to get 3 heads in this example, but now it is one way out of 4 possibilities. The probability of getting 3 heads once I've gotten 1 head is therefore 1/4.

Generalizing, let's let event A be getting a head on the first flip, and let event B be getting 3 heads after all 3 flips. The probability of getting 3 heads

TABLE 10.3. Three-Coin Conditional Probability Example

	Coin	
A	B	C
H	H	H
H	H	T
H	T	H
H	T	T
T	H	H
T	H	T
T	T	H
T	T	T

once I've gotten one head is the probability of B given A, written as $P(B|A)$. This must be equal to the probability of both events happening (B and A) divided by the probability of getting A. That is,

$$P(B|A) = \frac{P(B \text{ and } A)}{P(A)} = \frac{1/8}{1/2} = \frac{1}{4}$$

This same equation is often written slightly differently. Multiplying both sides by $P(A)$, I get

$$P(B \text{ and } A) = P(A)P(B|A)$$

This is known as the *multiplication rule*. If B doesn't depend at all on A—for example, the probability of a coin flip yielding a head after another coin flip yielded a head—then $P(B|A)$ is simply $P(B)$, and the above formula becomes the familiar simple multiplication rule for *independent* (i.e., not conditional) events,

$$P(B \text{ and } A) = P(A)P(B)$$

Note that the general multiplication rule doesn't know which event you're calling A and which you're calling B, so that

$$P(B \text{ and } A) = P(A)P(B|A) = P(A \text{ and } B) = P(B)P(B|A)$$

Consider the rolling of two dice. Each die is independent of the other so we might not expect the above conditional probability relationships to be needed. There are, however, situations where we define a conditional relationship into the question. For example, suppose we want to know what the probability is of the sum of the two dice equaling 5 when we roll the dice sequentially and the first die gives us 3 or less.

Let A be the first event, where we roll one die and get 3 or less. Since getting 3 or less must happen half the time, this probability is 0.5. Let event B be the second event, where we roll the second die and get a total of 5.

$P(A \text{ and } B)$ is the probability of getting a total of 5 from rolling two dice while the first die is 3 or less. The possibilities are {1, 4}, {2, 3}, and {3, 2}, giving us a probability of 3/36.

The probability of getting a total of 5 when the first die gave us 3 or less is therefore

$$P(B|A) = \frac{P(A \text{ and } B)}{P(A)} = \frac{3/36}{1/2} = \frac{1}{6}$$

For simple problems it is often easiest to list the possibilities, show the reduced event space, and get the results. Table 10.4 shows just such a situation.

We have a bag with 2 red marbles and 2 green marbles. We want to draw them out, one at a time. The table shows every possible order of drawing these 4 marbles. The marbles are labeled R1 and R2 for the two red marbles, G1 and G2 for the two green marbles. Since there are 4 choices for the first draw, 3 choices for the second draw, 2 choices for the third draw, and 1 remaining choice for the fourth draw, there are 4! = 24 possible combinations, as shown. Of these 24 choices, 4 of them have the 2 red marbles being chosen first. The probability of doing this is therefore 4/24 = 1/6.

I'd like to know the probability of drawing a second red marble once I've drawn a red marble. If I call A the event of drawing a red marble first and B the event of drawing the second red marble once I've drawn a red marble, then I'm looking for $P(B|A)$.

From above, $P(AB) = 1/6$. Since there is an equal number of red and green marbles before I start, the probability of drawing a red marble first is 1/2. $P(A)$, therefore, is 1/2. Thus,

TABLE 10.4. Four-Marble Conditional Probability Example

Draw from Bag in Order:			
R1	R2	G1	G2
R1	R2	G2	G1
R1	G1	R2	G2
R1	G1	G2	R2
R1	G2	R2	G1
R1	G2	G1	R2
R2	G1	R1	G2
R2	G1	G2	R1
R2	G2	R1	G1
R2	G2	R1	G1
R2	R1	G1	G2
R2	R1	G2	G1
G1	G2	R1	R2
G1	G2	R2	R1
G1	R1	R2	G2
G1	R1	G2	R2
G1	R2	R1	G2
G1	R2	G2	R1
G2	R1	R2	G1
G2	R1	G1	R2
G2	R2	R1	G1
G2	R2	G1	R1
G2	G1	R1	R2
G2	G1	R2	R1

$$P(B|A) = \frac{P(AB)}{P(A)} = \frac{1/6}{1/2} = \frac{2}{6} = \frac{1}{3}$$

Doing this problem "manually," the probability of drawing a red marble first is 1/2. Once I've drawn a red marble, there is 1 red marble and 2 green marbles left (the reduced event space), so the probability of drawing the second red marble is simply 1/3.

MEDICAL TEST RESULTS

A somewhat complicated but very important problem is the medical diagnosis problem. There is a deadly disease that afflicts, say, 0.1% of the population (one person in a thousand). It is very important to determine if you have this disease before symptoms present themselves. Fortunately, there is a diagnostic test available. This test is not perfect, however. It has two failings. The test will only correctly identify someone with the disease (give a positive result) 98% of the time. Also, the test will give a false-positive result (identify the disease in someone who doesn't have it) 2% of the time. You take the test and the results come back positive. What's the probability that you actually have the disease?

The answer to this question is important in cases where there is a vaccine available, but it has side effects. Let's say the vaccine causes a deadly reaction in 15% of the people who take it. If you really have a deadly disease, then a vaccine that could cure you is probably worth the 15% risk. On the other hand, if you really don't have the disease, then why take any risk with a vaccine? You would like to know what the probability of you having this disease is.

Let's start by defining the two events at hand:

D is the event that you have the disease.
T is the event that you test positive for the disease.

In terms of the information given above, we know that

$P(D)$ = the probability that you have the disease = 0.001.
$P(T|D)$ = the probability of a positive test given that you have the disease = 0.98.
$P(T|not\ D)$ = the probability of a positive test given if do not have the disease = 0.02.

We want to calculate two numbers:

$P(D|T)$ = the probability that you have the disease given that the test was positive.

$P(not\ D|not\ T)$ = the probability that you don't have the disease given that the test was negative.

We know that

$$P(D|T) = \frac{P(D\ and\ T)}{P(T)}$$

We can calculate the numerator directly from the known information, because we also know that

$$P(D\ and\ T) = P(D|T)P(T) = (.01)(.98) = .0098$$

Getting $P(T)$ requires a little manipulation. Calculate the intermediate number

$$P(not\ D\ and\ T) = P(T|not\ D)P(not\ D) = P(T|not\ D)(1 - P(D))$$
$$= (.02)(1 - .001) = .01998$$

and then

$$P(T) = P(D\ and\ T) + P(not\ D\ and\ T) = .0098 + .01998 = .02096$$

and finally,

$$P(D|T) = \frac{P(D\ and\ T)}{P(T)} = \frac{.00098}{.02096} = .0468$$

You have, according to these calculations, less than a 5% probability of actually having this disease. Intuitively, this doesn't seem reasonable: How can you test positive on a test that "catches" this disease 98% of the time and still have less than a 5% probability of actually having this disease?

The calculation is correct. The intuition issue comes from how I stated the information above. I left out a critical factor—the background occurrence of the disease. Let me state things a little differently, and also show a very quick and dirty way to estimate an answer: On the average, out of ten thousand people, ten people will have the disease (0.1% of 10,000 = 10) while about 200 people (2% of 10,000) will test positive for the disease. The probability of someone who tests positive actually having the disease is therefore about 10/200 = 5%. The reason why this is an approximation is that this simple estimate does not correctly take into account the overlap of people who test positive and those who actually have the disease. As was shown above, correctly handling this overlap drops the probability a bit from the estimate.

Armed with this probability, you are now in a much better position to decide whether or not to take a vaccine that has a 15% probability of killing you.

Figure 10.1 shows the probability that you have the disease, given a positive result on this same test, versus the background incidence of the disease $P(D)$ in the calculations above. Note that the horizontal axis in this figure is different than the horizontal axis in any previous figure. It is not laid out like a ruler. That is, the distance from 0.1 to 0.2 is not the same as the distance from 0.3 to 0.4, and so on. Instead the spacing is *logarithmic*. That is, there is equal spacing for multiples of 10: The distance from 0.01 to 0.10 is the same as the distance from 0.10 to 1.0, and so on. This is done here strictly to make the graph easier to study. If this axis had been laid out with equal distance for equal numbers (call *linear* spacing), then the information for the range below $x = .001$ would have been squeezed up against the vertical axis and would have been very hard to study. Strictly speaking, since the vertical axis spacings are linear and the horizontal axis spacings are logarithmic, this type of graph is called a *semilog(arithmic)* graph.

Going back to Figure 10.1, you can see that at the extremes, it almost doesn't matter whether or not you even took the test. If almost nobody has the disease (the left-hand side of the graph), then if it's very unlikely that you have the disease. An essentially flawless test is required to give you useful information. At the other extreme, if almost everybody has the disease, then

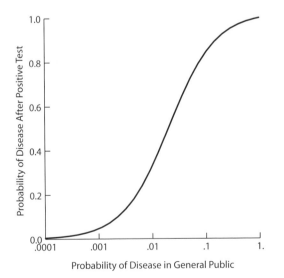

Figure 10.1. Probability of having disease after positive test versus probability of disease in the general public. Test false-positive rate = 2%, test false-negative rate = 2%.

you probably also have the disease. The test as it stands will have a high probability of being correct, but the information you gain doesn't add much to what you already know. For the ranges in between the two extremes, the test gets more accurate as the background (general public) incidence of the disease increases. If 1% of the public has the disease, the test is pretty useful. If 10% of the public has the disease, the test is very accurate.

I haven't yet finished the calculations for this example. I still need to calculate $P(not\ D|not\ T)$, the probability that you don't have the disease given that the test result was negative. This is simpler than the previous calculation because so many of the intermediate numbers are already available. First,

$$P(not\ T) = 1 - P(T) = .97903$$

and

$$P(not\ D) = 1 - P(D) = .99900$$

Then

$$P(not\ D\ and\ not\ T) = P(not\ D) - P(not\ D\ and\ T) = .99900 - .01998 = .97902$$

and finally

$$P(not\ D|\ Not\ T) = \frac{P(not\ D\ and\ not\ T)}{P(not\ T)} = \frac{.97902}{.97903} = .99999$$

This is as good as it gets. If the test came back negative, you don't have the disease.

All of the calculations above can easily be done, quite accurately, on a hand calculator. I should say a word of warning, however, about not rounding off numbers. Since we are looking at both the ratios of two similar numbers and the differences between two similar numbers, it is very easy to get things nonsensically wrong. For example, $1.001 - 0.999 - .001 = 0.001$. This answer could be important if it has to do with something like the death rate due to a disease. If the original numbers are too casually rounded off three significant figures, however, we would have $1.00 - 1.00 - .001 = -.001$ which in terms of a death rate makes no physical sense whatsoever.

THE SHARED BIRTHDAY PROBLEM

The last example I'd like to present on the topic of conditional probabilities is one of the most popular probability book examples. Given a group of n people, what is the probability that at least two of these people share a birthday? I'll consider only 365-day years and a group with randomly distributed birthdays (no identical quadruplets and the like). I could do this by formal conditional probability calculations, but this problem—like so many popular probability brain-teaser problems—is best solved by a trick. In this case it is much easier to solve the problem of the probability that in a group of n people no two people share a birthday. Then all I need is to subtract this result from 1 and I'm done. This type of calculation was used earlier to calculate the probability of repeated digits in a group of 10 randomly chosen digits.

Pick the first person in the group, an arbitrary choice, and note his/her birthday. Then pick the second person, again an arbitrary choice. The probability that this latter person's birthday is not on the same date as the first person's birthday is $364/365 = 0.99726$. The probability that both of these people share a birthday is therefore $1 - 0.99726 = 0.00274$.

Moving on to a third person, if the first two people don't share a birthday, then there are two days on the calendar marked off. The probability that this third person doesn't share a birthday with either of the first two is therefore $363/365$ and the probability that there is no match between any of the three is $\dfrac{364}{365}\dfrac{363}{365} = 0.99180$. The probability that at least two of these three people share a birthday is then $1 - 0.99180 = 0.0082$.

Continuing on (and this is very easy to do on a spreadsheet or even with a hand calculator, using the memory functions) is very straightforward. Just keep repeating the multiplication of decreasing fractions until you've included as many people as you want, then subtract the result from 1. Figure 10.2 shows the result of these calculations for n up to 70 people. Note that the probability

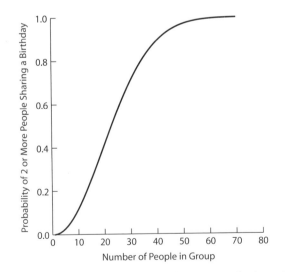

Figure 10.2. Probability of two or more people in a group sharing a birthday versus the size of the group.

of at least two people sharing a birthday crosses 50% at 23 people and reaches 99.9% at 70 people.

Conditional probability calculations show up often in real-world situations (such as the medical test for a disease) because in the real world so many situations have "strings attached." What I called the "multiplication rule" above is more often called "Bayes Rule" in honor of the man who developed it.

CHAPTER 11

SCHEDULING AND WAITING

Unfortunately, many of our daily activities involve waiting. We have to wait for the bus, we have to wait for to see our doctor in her office, we have to wait when we meet a friend for lunch, and so on. Wouldn't it be nice to have your bus pull up just a minute after you reach the bus stop, or have your doctor take you in for your examination just a minute after you reached her office? On the bus company's side, there is a trade-off between minimizing the number of buses (operating cost) yet not overloading buses and not having people wait too long in possibly inclement weather. On the doctor's side, there are clear advantages to getting to see as many patients as possible in a day while being able to go home for dinner at a regular time (and not having patients waiting long times for treatment). When you meet a friend for lunch, the exact time of the meeting usually isn't critical, but if one of you is 15 minutes early and the other is 15 minutes late, then one of you will just have to wait a while and ponder the value of timeliness.

I'll start with the doctor's office and see what we can learn, then I'll look at meeting friends and then bus schedules. My principal tool will be the Monte Carlo simulation, so in many cases I'll have to quote results without showing exactly how I got these results. The explanation follows directly from the discussion of Monte Carlo simulations, I just set up the situation and ran about a million cases.

Probably Not: Future Prediction Using Probability and Statistical Inference,
by Lawrence N. Dworsky.
Copyright © 2008 John Wiley & Sons, Inc.

SCHEDULING APPOINTMENTS IN THE DOCTOR'S OFFICE

I'll assume that the working day is from 9 to 5. Even though real people like to break for lunch, I'll assume that our doctors work straight through. I'm doing this because taking account of lunch breaks complicates the calculations and conclusions without adding anything to our understanding of the statistics of the situation(s). Also, I'm going to ignore the possibility that someone calls at the last minute with a medical need that requires at least prompt, if not immediate, attention. This can happen to a physician, and the best she can do is to juggle the remaining appointments and try to notify everybody of the changes before they commit to showing up at the office.

The simplest situation requires no detailed analysis. If you know that a visit takes 1/2 hour, then you stack up 16 appointments and your day is perfectly filled. Let your patients know what time their appointment is, and none of their time is wasted. Life couldn't be better.

If different visits take different amounts of time but you know exactly how long each visit will take, then you have to spend a few minutes writing things out and adding up the times, but you can still do pretty well. The only problem is if, for example, six people call for appointments for visits that take 1.5 hours each. If you just schedule five of these appointments for the day, then you're finished at 4:30 and haven't efficiently filled your day. If you try to fit in all six people, then you're not going home at 5:00.

If you get requests for appointments for visits of differing amounts of time, so long at you know each visit time, you can optimize things reasonably well with just a pencil and paper. What do you do, however, if you have no a priori idea how long each visit will take? While in truth you probably can estimate required times to some extent based on the complaint accompanying the request for an appointment, let's start with an extreme case: Visits will vary randomly in required time uniformly from 15 minutes to 2 hours, and you have no idea how long each visit will take until you've actually seen that patient.

The first observation is that if you want to be absolutely sure that you can leave by 5:00, you cannot make any appointments for later than 3:00. Since all the visits might be 2 hours long, you can only schedule four appointments a day. The Monte Carlo simulation, however, shows that if you schedule four appointments each day, then on the average you'll be going home at about 1:30 with a standard deviation of about 1 hour. This certainly doesn't look like a good way to do things.

Scheduling five appointments a day, your average day will end at about 2:45 with a standard deviation of about an hour and 10 minutes. There's still a lot of wasted afternoon time, but we have to look at the distribution if we want to be very sure that you'll never have to stay after 5:00.

Figure 11.1 shows the distribution of working day lengths. I subtracted the 9:00 starting time from the data—it would just represent a horizontal offset on the graph. The graph looks normal. Since we're only interested in the cases where you'd work past 5:00 (8-hour working day), to get a 95% confidence

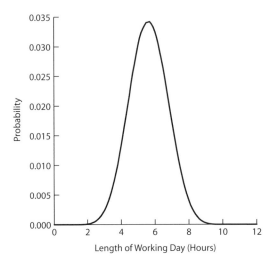

Figure 11.1. Distribution of the length of a working day.

factor we need 1.65 standard deviations, which comes out to about 4:45. This means that scheduling five appointments a day gives you a 95% confidence factor that you'll be able to leave by 5:00. This also means that about 5% of the time you'll work past 5:00, but all in all, this looks like a pretty good situation that meets the specified conditions almost all the time.

The three "not so great" consequences of scheduling five appointments a day are that you'll end up leaving work early most of the time (maybe you could take up golf), you're not seeing as many patients as you might because of this wasted time, and if all your patients show up at 9:00 in the morning, most of them are going to be sitting around for a long time. This latter situation can be helped a little bit: Since the minimum appointment time is 15 minutes (1/4 hour), there's no way that you could see the second patient before 9:15, the third patient before 9:30, and so on, so you could give them a little relief in their appointment times. Let's try and quantify some of this.

The average amount of wasted time is 2:20. This is a lot of wasted time. Note that the average go-home time plus the average wasted time comes out to a little more than 8 hours. This is not a mistake. If you take this sum and subtract the amount of time that's spent after 5:00 on the occasional days that working late is necessary, you'll get 8 hours. This is a case where the mathematics and the psychology don't line up: Mathematically, if time before 5:00 spent not working is wasted, then time spend after 5:00 working must be called "negative wasted time."

Now, what about the unfortunate patients who all showed up at 9:00 but only one (each day) gets seen immediately? Table 11.1 shows the average waiting time and the standard deviation of these waiting times for each of the five appointments of the day. Note that I've switched from hours and minutes

TABLE 11.1. Average Waiting Times for Each of Five Appointments

Appointment Number	Average Wait (hours)	Sigma Wait (hours)
1	0	0
2	1.13	0.500
3	2.25	0.710
4	3.37	0.873
5	4.50	1.007

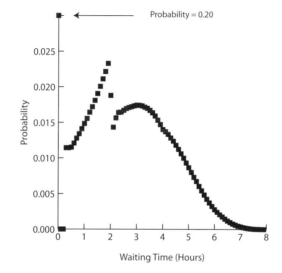

Figure 11.2. PDF of patient waiting time (simple scheduling).

to hours and decimal parts of an hour. It's not the way we usually tell time(s), but in this context I think it's easier to follow.

The person who gets the first appointment is lucky—there's no wait at all. With each succeeding appointment, however, the story just gets worse. Not only does the average wait time increase, but the standard deviation also increases. The patient with appointment #5 can expect to wait four and a half hours, with a standard deviation of 1 hour!

The preceding paragraph is mathematically correct, but doesn't really tell the whole story. A patient showing up at 9:00 doesn't know which appointment number they will get. Looking at all the wait times for all the appointments, the average wait is 2.2 hours, with a standard deviation of 1.7 hours.

Figure 11.2 shows the PDF of the waiting times. This is indeed an odd-looking graph. The first point (probability = 0.2) shows that 1 out of 5 of the patients will have no wait at all. I've distorted the probability (vertical) axis to show this point because if I ran this axis all the way up to 0.2, the other points would be so close to the horizontal axis that you wouldn't be able to see them clearly.

All in all, this is a pretty pitiful situation. I've exaggerated my assumptions in that while it's possible to have visits as short as 15 minutes or as long as 2 hours, these will occur infrequently as compared to 30-minute to 1.5-hour visits. In other words, the visit times between 15 minutes and 2 hours will not all be equally likely.

Let's see if I can improve the situation by adding a little bit of intelligence to the scheduling process. The nurse or receptionist making the appointments has had enough experience to take a guess at the necessary visit time when writing down the response to "What seems to be bothering you?" It's not a very accurate guess—let's say the accuracy can be represented by a normal distribution with a sigma of 25% of the mean (the guess).

The office will run this way. When a call comes in requesting an appointment, the receptionist will estimate the length of the visit in half-hour increments, from 0.5 hours to 1.5 hours. In the simulation, these appointments will be uniformly distributed random choices. The actual length of the visit will be a random number chosen from a normal distribution with the estimate as the mean and 25% of the estimate as the standard deviation. Several lists of eight appointments and visits are shown in Table 11.2. This isn't too different from the original choice of appointment times, but in this case I'm acknowledging that the very short and very long appointments will be less usual than the more "middle of the road" appointments.

I chose eight appointments in a day for this table because with a uniform distribution of 0.5-, 1.0-, and 1.5-hour appointments, the average day should contain 8 hours of appointments. Unfortunately, very few days are average. For the 3 days shown in the table, my computer spit out total appointment times of 6.0, 8.0, and 8.5 hours. The corresponding simulated visit times came out as 6.0, 8.6, and 7.1 hours, respectively. If we want to keep your "I want to be out of the office by 5:00 at least 95% of the time" rule, I'll have to back off from trying to fit eight appointments into the day. Also, I'll have to base the simulations on the visit times, not the appointment times.

Table 11.3 shows the standard deviations of the appointment and visit times for different numbers of scheduled appointments in the day.

I'll keep the same criterion that I used in the first examples, that I want the average +1.65 standard deviations of the total visit times for the day to be less than 8 hours. For five appointments per day, this calculation is $5 + (1.65)(1.09) = 6.81$ hours. For six appointments per day, I get $6 + (1.65)(1.20) = 7.98$. This is very good news. Just by adding a little intelligence to the scheduling process, I am able to increase the number of appointments from five to six and still be 95% confident of leaving the office by 5:00.

Now, what about the patients' waiting times? I could perform the same calculations as above and report the results, but instead I'd rather take a very big jump: Rather than asking all the patients to come in at 9:00, why not take

TABLE 11.2. Three Samples of Estimated (Appointment) and "Actual" Visit Times

Appointment Number	Scheduled Wait (hours)	Actual Wait (hours)
1	0.5	0.49
2	1.5	2.55
3	0.5	0.66
4	0.5	0.41
5	1.0	1.15
6	0.5	0.30
7	0.5	0.60
8	1.0	0.68
1	1.0	0.54
2	1.5	1.42
3	1.5	1.88
4	1.0	1.40
5	0.5	0.54
6	0.5	0.56
7	0.5	0.45
8	1.5	1.79
1	1.0	1.02
2	1.0	0.84
3	1.5	1.39
4	1.5	0.48
5	0.5	0.56
6	0.5	0.62
7	1.0	1.19
8	1.5	0.99

TABLE 11.3. Standard Deviation of Total Days' Appointment and Visit Times

Number of Appointments	Sigma Appointments	Sigma Visits
4	0.82	0.98
5	0.91	1.09
6	1.00	1.20
7	1.08	1.29
8	1.15	1.38

advantage of information we have about how the day is going to "play out?" I want to be very sure that there's a patient waiting in the office when his appointment comes up, so I'll tell each patient his appointment time minus three standard deviations.

TABLE 11.4. Three Samples of Appointment, Visit, Arrival, and Wait Times

Appointment Number	Appointment Time	Actual Time	Arrival Time	Wait Time
1	9.00	9.00	9.00	0.00
2	9.50	9.49	9.13	0.36
3	11.00	12.04	9.81	2.22
4	11.50	12.69	10.26	2.44
5	12.00	13.11	10.70	2.40
6	13.00	14.26	11.50	2.76
1	9.00	9.00	9.00	0.00
2	9.50	9.60	9.13	0.48
3	10.50	10.28	9.66	0.62
4	11.50	10.83	10.38	0.45
5	13.00	12.25	11.41	0.84
6	14.50	14.13	12.55	1.58
1	9.00	9.00	9.00	0.00
2	9.50	9.54	9.13	0.41
3	10.00	10.09	9.47	0.62
4	10.50	10.55	9.85	0.70
5	12.00	12.34	10.70	1.64
6	13.00	13.36	11.50	1.86

Table 11.4 shows a few examples of the simulation. This is beginning to look like a real physician's office in that if you can get the first appointment of the day, you have no wait. If you get an early morning appointment, you can expect a short wait. If you get a late afternoon appointment, then you'll probably have a significant wait.

Overall, has the wait situation improved? The average wait for all patients is now 1.1 hours with a standard deviation of 0.78 hours. Both of these numbers are approximately half what they were in the first example. Figure 11.3 shows the PDF for the waiting times. Again, the value at 0 waiting time is shown out-of-scale just to make the rest of the curve easier to read. This curve looks similar to Figure 11.2, as it should. Since we're now adding up 6 numbers rather than 5, Figure 11.3 (aside from the "spike" at 0 wait time) looks a bit more normal than does its predecessor.

Note that there are a few values of negative wait times. This is because I only required that patients arrive 3 standard deviations before their estimated appointment time. These negative values represent times when the doctor is ready to see the next patient, but the next patient hasn't arrived yet. This only happens about 1/20% of the time, so it doesn't influence the statistics meaningfully.

We really have here a success story. By just adding a little intelligence to the scheduling process, the number of patients the doctor sees in a day has

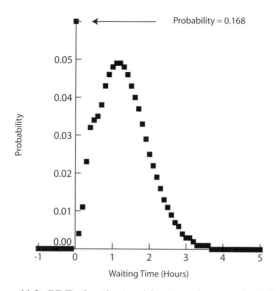

Figure 11.3. PDF of patient waiting time (smart scheduling).

gone from 5 to 6, the average patient waiting time has dropped in half, and the standard deviation of that average has dropped in half. In corporate English this is known as a "win–win" situation.

If a system such as this is implemented, there is an opportunity for continuous improvement. The scheduler should keep records of actual visit times versus the estimates and, by reviewing the comparisons, learn to improve the estimates—possibly going to a different number of "standard appointment lengths" and certainly tightening up on the standard deviations by better estimating the correct appointment length.

LUNCH WITH A FRIEND

In Chapter 3, I looked at the sums of two random numbers. But what about the differences between two random numbers? This is actually not something new, because of basic algebraic relationships. The subtraction 3 − 2, for example, is equivalent to the addition 3 + (−2). That is, instead of subtracting two positive numbers I'm adding a positive number and a negative number. The results of course are the same for both cases.

Why am I bringing this up now? Because this is the way to attack problems of waiting. For example, suppose you and I are to meet at a restaurant at 1:00 for lunch. You tend to be early. Specifically, you'll show up anywhere from on time to 15 minutes early, with a uniformly distributed probability of just when you'll show up. I, on the other hand, tend to be late. I'll show up any time from on time to 15 minutes late, again with a uniformly distributed probability of just when I'll show up. It isn't really critical just what time we start lunch, but there's that annoying gap between your arrival and my arrival when you're wondering just how much my friendship is really worth.

Let's take the nominal meeting time as our reference. Your arrival time is then a uniformly distributed random variable between 0 and 15 minutes and my arrival time is a uniformly distributed random variable between 0 and −15 minutes. The PDF of the time gap between our arrivals is the difference between your arrival time and my arrival time. I could work this out step by step or you could go back to Building a Bell (Chapter 3) and just look at the first example again. The resulting PDF is shown in Figure 11.4.

The minimum gap time is 0, which happens when we both show up exactly on time. The maximum gap time is 30 minutes, which happens when I'm 15 minutes late and you're 15 minutes early. The expected value, which is equal to the average because the curve is symmetric, is just 15 minutes. The standard deviation of each of our arrival PDFs is 6.1 minutes, and since our arrival times are uncorrelated, the standard deviation of the gap time (Figure 11.4) is just

$$\sigma = \sqrt{6.1^2 + 6.1^2} = \sqrt{2(6.1)^2} = 6.1\sqrt{2} = 6.1(1.414) = 8.66 \,(\text{minutes})$$

Now let's change things a little bit. I'll make us both pretty good citizens in that on the average we both get to our appointments on time. However, we both are not perfect in that we will arrive some time within 7.5 minutes early and 7.5 minutes late. Again, I'll make the PDFs uniform.

The PDF for my arrival time is the same as the PDF for your arrival time, as shown in Figure 11.5. I've referenced everything to an ideal arrival time of

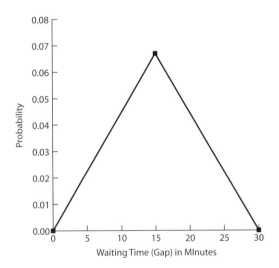

Figure 11.4. PDF of gap between arrival times when you're always early and I'm always late.

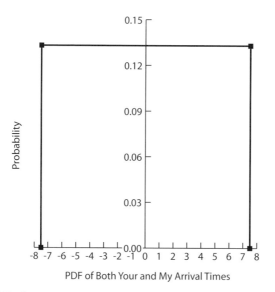

PDF of Both Your and My Arrival Times

Figure 11.5. PDF of your and my arrival times when we're both nominally on time.

0 so that early arrivals are represented by positive numbers and late arrivals are represented by negative numbers.

The time that one of us will have to wait for the other is just the difference between your arrival time and my arrival time. I have to subtract one PDF from the other. At the beginning of this section, I noted that subtracting a PDF is the same as adding the negative of the PDF. For a PDF that's symmetric about (with an average of) zero, however, the negative of the PDF is just the PDF again. In other words, there's no difference between adding and subtracting two of the PDFs shown in Figure 11.5.

The result of combining these two PDFs is shown in Figure 11.6. The standard deviation is the same as that of Figure 11.4, namely, 8.66.

Figure 11.6 is mathematically correct and shows that the expected value of the difference between our arrival times is 0. Since we both, on the average, show up on time, this is not a surprise. However, Figure 11.6 doesn't show what is probably the most useful information for this situation—that is, how long should one of us expect to be sitting in the restaurant alone, waiting for the other?

If we just want to answer this question, then what we want is not the difference between the two arrival times but rather the absolute value of the difference between the two arrival times. That is, if I arrive 10 minutes early and you arrive 5 minutes late, the waiting time is 15 minutes. If you arrive 10 minutes early and I arrive 5 minutes late, the waiting time is still 15 minutes. There is no such thing as a negative waiting time in this sense.

Figure 11.7 shows the PDF for this new definition of waiting time. The minimum waiting time is 0, which occurs when we both show up at the same

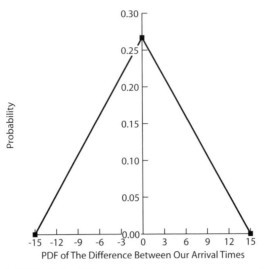

Figure 11.6. PDF of the difference between your and my arrival times.

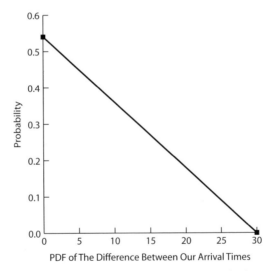

Figure 11.7. PDF of the waiting time between luncheon arrivals—uniform lateness.

time, whatever that time might be. The maximum time is 30, which occurs when one of us is 15 minutes early and the other of us is 15 minutes late. It doesn't matter which of us does which. The expected value of the waiting time is 5 minutes and the standard deviation is 3.5 minutes.

Next, I'll replace the uniform distribution functions with the more reasonable normal distribution functions. Again, we both tend to show up on time,

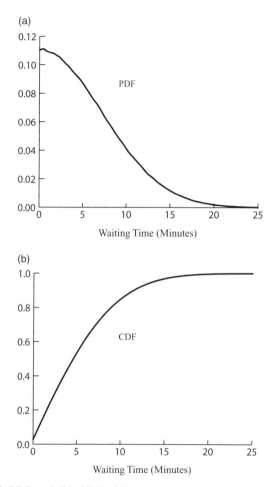

Figure 11.8. (a) PDF and (b) CDF of the waiting time between luncheon arrivals—normal lateness.

but our actual arrival times are normally distributed with a standard deviation of 5 minutes.

The resulting PDF looks like the right half of a normal distribution (Figure 11.8a). The expected wait time is 5.62 minutes and the standard deviation is 4.22 minutes. Remember that the right half of a normal distribution is not itself a normal distribution, so be careful about how you'd calculate a confidence interval. The CDF is shown in Figure 11.8b. This is a one-sided confidence interval, waiting less than 0 (minutes) is meaningless. The 95% confidence interval is from 0 to ~14 minutes.

Fortunately, most good restaurants have bars to wait in.

WAITING FOR A BUS

Waiting for a bus is different from waiting to meet a friend. Waiting for a friend is symmetric in the sense that it doesn't matter if you or your friend got to the meeting place first. The situation is the same if you get to the bus stop before the bus and have to wait for your "meeting." On the other hand, if the bus gets to the bus stop before you do, there won't be any meeting—you'll have missed the bus. If there is only one bus that day and you have to get to your destination that day, then it is critical that you don't miss the bus. You have to look at the statistics of when the bus arrives as compared to the published schedule, and then you have to be honest with yourself about how well you meet a schedule. If we assume that both events are normally distributed with a mean of the time the bus schedule says the bus will arrive, then your target time for meeting the bus (with a 97.5% confidence factor) must be earlier than the bus schedule time by $2\sqrt{\sigma_{you}^2 + \sigma_{bus}^2}$. If you need a better confidence factor, then you must target getting to the bus station even earlier.

An entirely different problem arises from the situation where there is more than one bus coming past your stop each day. Let's assume that buses run hourly. In this case (except for the last bus of the evening), missing a bus can be reinterpreted as having 100% probability of catching the next bus. The interesting part of the problem is looking at what wait times will be and if there is any optimization (i.e. minimization) of your wait time to be had.

There are so many possible variants of this situation that it's impossible to consider all of them. I'll pick a few of them and see what happens.

Suppose that buses are supposed to come hourly, on the half-hour. There's a 1:30 bus, a 2:30 bus, and so on. However, due to varying traffic conditions, weather, road construction, and so on, the buses never arrive when they're supposed to. The 1:30 bus will arrive some time between 1:00 and 2:00, the 2:30 bus will arrive some time between 2:00 and 3:00, and so on. You want to catch the 1:30 bus. Since you've got no idea just when the bus will arrive, you get to the bus stop at some time between 1:00 and 1:59. What can you expect in terms of wait times and which bus you'll actually catch?

Assuming all three PDFs (two buses & you) are uniformly distributed, there's a 50% probability that you catch the first bus and a 50% probability that you'll catch the second bus. Your shortest possible wait time is zero; this happens when the first bus pulls up immediately after you get to the bus stop. This can happen any time between 1:00 and 2:00. Your longest possible wait time is just under 2 hours—the first bus comes at 1:00, you arrive at 1:01, and the second bus comes at 2:59.

Figure 11.9 shows the PDF and the CDF for waiting times. This is a true continuous distribution, so it must be interpreted carefully. We cannot say, for example, that there is a probability of 1.0 of zero wait time, even though the graph sort of *looks* that way. With a continuous PDF, we calculate the probability of an occurrence over a range of the random variable by calculat-

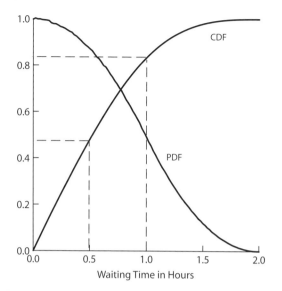

Figure 11.9. PDF and CDF of waiting times for a bus with all random arrivals.

ing the area under the curve for this range. Usually this takes integral calculus, and the CDF provides the results of these calculations. For example, the probability that you'll have to wait between 0 and 0.5 hours is a little less than 50%. The probability that you'll have to wait between 0 and 1 hour is about 85%. The probability that you'll have to wait between 0.5 hours and 1 hour is about 35% because 85% − 50% = 35%. Also, while the PDF might look like the right-hand side of a normal curve, it most certainly is not (part of) a normal curve. This PDF goes to 0 at wait time = 2 hours, whereas a normal curve would slowly approach zero forever.

The PDF shows that it is much more probable that you'll have a short wait than a long wait.

In order to see if there is a best time to get to the bus stop, I changed your arrival time at the bus stop from a random variable to a fixed time. Figure 11.10 shows the expected wait time and standard deviation of the expected wait time for arrival times from 1:00 to 2:00 in 1/10-hour (6-minute) intervals. Both the expected wait time and standard deviation for arrival times of 1:00 and 2:00 are the same. This is because the situation at both of these times is identical: You arrived at the beginning of the hour with a bus definitely coming at some time during that hour. The worst time to come is at 1:30 because this time has the longest expected value of waiting time. This was, at least to me, an unexpected result—the worst time to show up is the time on the bus schedule! However, note that slightly before 1:30 while the waiting time drops, the standard deviation of this waiting climbs (both as compared to 1:30). All in all, it looks like you want to come just before the hour—1:00 if you want to be sure to catch the first bus, 2:00 if you're OK with the 2:00 bus.

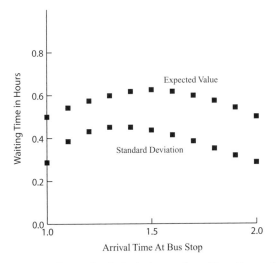

Figure 11.10. Average and standard deviations of waiting times for fixed arrival times.

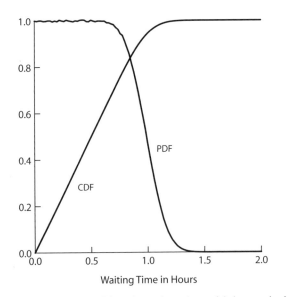

Figure 11.11. PDF and CDF of waiting times for a bus with bus arrivals normally distributed and rider arrivals uniformly distributed.

As a last exercise, I'll consider how things actually usually work. The buses are supposed to arrive every hour on the half-hour. They nominally meet this schedule, but with a normal distribution centered around the half hour. I'll set a sigma of 0.1 hours (6 minutes). If you arrive at the bus stop 12 minutes before

1:30, you're 95% confident of catching the 1:30 bus. What happens if you just wander over to the bus stop (between 1:00 and 2:00) whenever the spirit moves you?

Figure 11.11 shows the result of this calculation. This figure doesn't look all that different from Figure 11.9. The expected waiting time is about 35 minutes with a standard deviation of 20 minutes.

The discussions in this chapter about meeting a bus and waiting in a doctor's office point to how much improvement in scheduling and waiting times can be had if there is good information about all the variabilities involved. The discussion about meeting a friend is really just an exercise in calculating means and standard deviations; the only improvement to be had is for both parties to show up as close to on time as possible. In terms of the bus and the doctor's office, however, good use of the PDFs involved show how some statistical information about the upcoming day's events can really be used to help make the day better.

CHAPTER 12

STOCK MARKET PORTFOLIOS

If you are at all interested in the stock market, no doubt you have at some time read some magazine's pick for the "Portfolio Manager of the Year" or the "Mutual Fund of the Year." If you're a little bit cynical, you've also read the descriptions of how throwing darts at a list of New York Stock Exchange stocks can be the basis of a portfolio that will outperform many of the "better" mutual funds. Or maybe you're cynical because you have read about these things.

I could write a Monte Carlo simulation in which I track, say, 100 different portfolios of, say, 50 stocks each and see how they do after three years. I could try imposing some rules such as dumping losing stocks every 3 months and replacing them with the best performing stocks of the past 3 months. I would run the simulation for a (simulated time of) a few years and compare the results to the last 3 years of the NYSE. When my simulation is complete, I could present the results and show that some portfolios did well and some did not do well.

This type of simulation has been done and written about many times. I don't think I have anything to add. Instead, let's see if I can add some insight into why the results come out the way they do.

I'll start my discussion with a very simple old game. I randomly choose 64 people scattered about the country. On the evening before the first game of

Probably Not: Future Prediction Using Probability and Statistical Inference,
by Lawrence N. Dworsky.
Copyright © 2008 John Wiley & Sons, Inc.

the world series, I telephone all of these people. The first 32 I tell that the American League team will win the game the next day, the remaining 32 I tell that the National league will win the game the next day. Clearly my prediction to 32 of these 64 people will be correct.

The next day I cross the names of the 32 people that I had given an incorrect prediction to off my list. On the evening before the second game, I call the 32 people that I had given a correct first game prediction to. I tell 16 of these people that the American League team will win the game the next day, and I tell the remaining 16 people that the National League will win the game the next day. Again, clearly my prediction to 16 of these 32 people will be correct.

You can see where I'm going. After 6 games of the World Series, there is one person out there who has received 6 correct predictions in a row about the next day's baseball game. On the evening before the seventh game, I telephone this person and request several thousand dollars for my prediction for the next day's game.

A variation of this game is to write down the 64 possible outcomes of the first six games any time before the start of the world series and mail one of these to each of 64 people. On the night before the seventh game, I simply telephone the person holding the sheet that has the correct results for the first six games. Mathematically, both of these approaches are identical; I'm not sure which one would have better psychological impact.

Suppose you are the person I call on the night before the seventh game. What would you do? If you hadn't been reading this or some other book about these sorts of things, you might believe that I really could make predictions, or had a system, or had a friend on a UFO that traveled through time, or something like that. If you had some knowledge of these sorts of calculations and scams, you might suspect what had been going on. Is there any way you can be sure?

Let me turn the situation around with another scenario. You are a financial planner or the manager of some mutual fund. You have a good education in the stock market and you believe that you know how to pick stocks. You sit down one evening and do your calculations, roll your dice, consult your Tarot cards, whatever, and come up with what you believe will be a winning portfolio for the upcoming year.

A year later, it turns out that you were right. Your portfolio performed very well. Your mutual fund was one of the better performers of the year. Your customers are very happy with you.

Now you sit down to plan your portfolio for year two. Over the course of year one, you have refined your theories and expanded your data files. You put together your portfolio for year two.

At the end of year two, it turns out that you were right again. The clients who have been with you for two years have parlayed their first year's earnings into second-year investments and even greater earnings. You are starting to be noticed by various financial magazines; you are interviewed for one of them, your picture and a small article about your successes appears in another, and so on.

Now you sit down to plan year three. As you start to work through your calculations, a terrible thought appears out of nowhere in your head and expands to almost explosive dimensions. Do you really know what you're doing or were you just lucky? Are you the mutual funds' equivalent of the guy who just won the lottery and believes that his system of picking "good" numbers really worked? What can you do to reassure yourself that it wasn't just luck?

I've now put forth two scenarios and ended them both with the same question. Did the person who has produced such good predictions and delivered them to you in a timely manner really know something more than the rest of us or was it just luck or some other ruse?

The answer is that you can never know for sure. Let me emphasize this point. There is no way of knowing for sure. The best you can ever do is to try and see if it's reasonable to assert that things went right just by chance.

In the case of the World Series scam, you can see that there needn't have been any fortunetelling ability involved. This situation is directly analogous to any anecdotal "sampling" of data. People who only get to see good results and don't even know that there are bad results out there are very easily fooled. If I could predict the result of all seven games before the first game is played and repeat this several years in a row, then either I'm telephoning an awful lot of people every night during the world series or I somehow am really predicting the future.

In the case of a lottery, there are no surprises. Hundreds of millions of lottery tickets are sold. Many, if not all, of the ticket purchasers believe that they have an approach or a "system." Unbelievably, there are books and newspaper articles about how to pick "good" numbers written by self-proclaimed "lottery experts."[1]

When there are approximately the same number of lottery tickets out there as there are possible winning numbers, it isn't hard to explain why there is a winner of the lottery. But what about the stock market? Just how unlikely is it that a portfolio picked by throwing darts at a list of stocks stuck on the wall will do well?

Simulating stock market performance is at best a slippery business. If all I'm interested in is a general look at how things seem to work and never fool myself into thinking that I can develop the computer program that will make me rich, however, then it's not too difficult to do some useful work.

I will be taking a very low-resolution look at stock market performance. For my purposes, stock prices on the first day of the year will be all I need.

---

[1] This begs the obvious question of why someone who knows how to pick winning lottery numbers is sitting around writing books and articles rather than simply exercising her skills and quickly surpassing Bill Gates and Warren Buffet combined in wealth.

Clearly this is not good enough for actual stock trading decisions, but again I'm only looking for a simple picture that makes sense.

If I divide a stock's price on January 2, 2006 by its price on January 2, 2005, I get a number that I can call the stock's *performance* for the year 2005. I'm ignoring dividends, splits, trading commissions, and so on, in this discussion. This simple calculation is equivalent to the interest I would have received had I put my money in the bank for a year rather than buying stock in a company. A performance number of 1.05, for example, means that I can get back $1.05 for every $1.00 invested if I sell my stock a year later. A performance number of 0.95, on the other hand, means that I would lose a nickel on every dollar invested. This performance calculation lets me compare stocks with different selling prices. This is necessary because a dollar per share earned on a stock that I bought for $100 a share is certainly not as impressive as a dollar per share earned on a stock that I bought for $3 a share.

If you randomly pick a couple of dozen stocks from, say, the New York Stock Exchange and calculate their performance numbers for the past few years (one year at a time) and then look at the distribution of these performance numbers, you'll get a distribution that looks pretty normal. This means that if you randomly choose a handful of stocks, you are sampling a distribution and can expect the performance of the average of your stocks to be the same as the average performance of all of the stocks in the population. If you were to buy one dollar's worth of every stock out there, then of course your portfolio's performance would be exactly the market's performance.

Let's say that the market's performance was flat for the year. That is, the average of the performance of all stocks is exactly 1.0. If you randomly picked, say, 25 companies and bought equal dollar amount numbers of shares of stock in these companies, then the expected value of the performance of your portfolio would be 1.0. If the standard deviation of all the stocks available is σ, then the standard deviation of the expected value of a large number of 25-stock portfolios built the same way you built yours (by random choice) would be

$$\frac{\sigma}{\sqrt{n}}$$

where n is the number of different companies' stock in your portfolio; in this case $n = 25$.

If the distribution of the population of stocks is approximately normal, then approximately the same number of stocks had performances greater than 1 as those that had performances less than 1. This means that there's a 50% chance that a randomly chosen group of stocks will do better than the average. The larger the number of stocks in the portfolio, the less probable it is that the average of the portfolio will differ significantly from the population average— but still, about 1/2 of the portfolios will do better than the average. About 1/4 of the portfolios will do better than the average 2 years in a row. The standard deviation of the sample is a measure of how wildly the performance of these small portfolios will vary.

The above logic also works for a year when the market does well or poorly. That is, about 50% of the portfolios will do better than average.

There's nothing new or exciting in what I said above. Basically I have just been reiterating the properties of a normal distribution. This isn't like the lottery where there is one big winner. There are a lot of portfolios "beating the market." There are so many beating the market that it's not hard to find some that not only are among the best performing of the portfolios but also repeat this accomplishment 2 or 3 years in a row. In other words, the fact that your carefully picked portfolio performed well for 2 years (assuming a flat or better market) isn't at all clear proof that you weren't just lucky. If the market fell during the year, then the probability that your results were just luck went down a little.

Now, I haven't given any reason for you to believe my assertion that my simple stock performance criterion does at all have a normal distribution when looking at a large population of stocks. It seemed reasonable to me, but I've fooled myself before. Therefore, I decided to take a look.

I obtained data for closing prices of all stocks on the New York Stock Exchange for the years 2001–2005. I then pruned this list to retain only those stocks that had daily listings over this entire period. I cannot claim that this pruning didn't distort my results. However, I'm only looking for a qualitative picture of stock performances here; and since after pruning I still had 1591 stock listings, I decided to charge ahead.

I calculated my performance numbers, as described above, using the closing price of the first day of trading for each of the years. This gives me 4 sets of yearly performance (2001, 2002, 2003, 2004).

Figures 12.1 through 12.4 show my results, with a "best-fit" normal distribution superimposed.[2] Once again I have connected the dots on the busy discrete data set just to make these figures easier to look at. As can be seen from looking at these figures, the NYSE stocks' yearly performance isn't really normally distributed, but the distributions are close enough to normal for the arguments made above to be reasonable.

In summary, if someone tells you he can pick a winning lottery ticket, against 100,000,000:1 odds, and then does it, pay attention. There's something going on here. On the other hand, if someone tells you that she can create a

---

[2] "Best Fit" is a procedure for matching a given continuous function to a set of data by minimizing the difference between the function and the data at the discrete points where the data exist. Typically an error function is created; this error function is the sum of the squares of the differences between each data point and the corresponding function value. In a normal distribution function the peak value, the mean, and the standard deviation are adjusted to minimize this error function. There are innumerable computer programs and spreadsheet functions for performing this task.

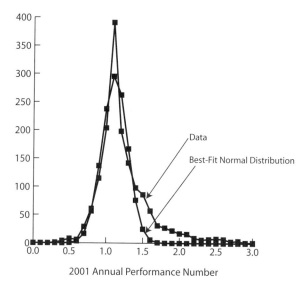

Figure 12.1. Approximate annual performance histogram of NYSE stocks for 2001.

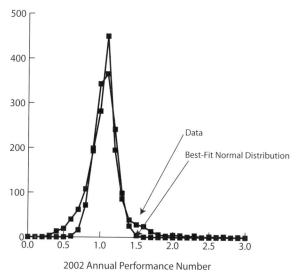

Figure 12.2. Approximate annual performance histogram of NYSE stocks for 2002.

stock portfolio that outperforms the market and most other portfolios and then does it a couple of years in a row, be interested but not yet convinced. She hasn't really yet demonstrated that she's working outside the laws of chance.

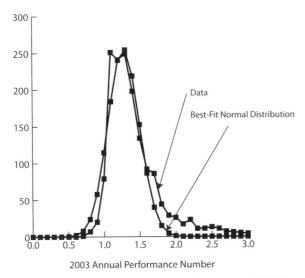

Figure 12.3. Approximate annual performance histogram of NYSE stocks for 2003.

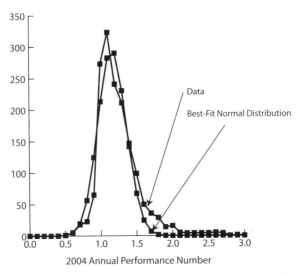

Figure 12.4. Approximate annual performance histogram of NYSE stocks for 2004.

This has been a very short chapter. From the perspective of stock prices as a collection of random walks, there simply isn't that much to say. There are of course many books written on how to play the stock market. Some of them might be very good. I wish I knew how to find out.

CHAPTER 13

BENFORD, PARRONDO, AND SIMPSON

This chapter will cover two unusual statistical anomalies, Benford's Law and Parrondo's Paradox, and will also cover one often misleading characteristic of averaging groups of data known as Simpson's Paradox. Benford's Law shows a distribution that seems to show up very often in nature and human endeavors, even though most people don't know it's there until they look very closely. Parrondo's[1] Paradox shows something that intuitively just seems wrong, but nonetheless is true—that it is possible to put together a gambling game out of two losing games (the odds are against you in both games) and come up with a winning game (the odds are with you). A case of Simpson's Paradox was shown in the Preface; there are quantifiable situations where combining data sets can lead to absolutely incorrect conclusions about some of the averages of the data. I'll try and explain what's happening and propose a procedure for getting correct results.

BENFORD'S LAW

Figure 13.1 shows the function $1/x$ for x between 1 and 10. I'll discuss the reasons for the choice of 1 and 10 a little later. For now just note that this

[1] Sometimes spelled Parrando's in the literature.

Probably Not: Future Prediction Using Probability and Statistical Inference,
by Lawrence N. Dworsky.
Copyright © 2008 John Wiley & Sons, Inc.

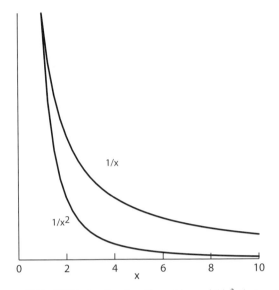

Figure 13.1. PDFs for the functions $1/x$ and $1/x^2$, $1 < x < 10$.

TABLE 13.1. Percentage of Leading Digits for Lists of Random Numbers Generated from the PDFs $1/x$ and $1/x^2$, with x between 1 and 10

	Fraction of Occurrences	
Leading Digit	$1/x$	$1/x^2$
1	.301	.555
2	.178	.185
3	.124	.093
4	.095	.056
5	.079	.037
6	.067	.027
7	.058	.020
8	.051	.015
9	.046	.012

function cannot be defined for $x = 0$, because $1/0$ is not a defined operation (a fancy way of saying that it makes no sense). Also shown in Figure 13.1 is the function $1/x^2$. This function is somewhat similar looking to $1/x$ and will be used to contrast properties.

Table 13.1 shows the results of counting the number of occurrences of the leading (leftmost, or most significant) digits in a large list (100,000 numbers) of random numbers generated from each of the functions above and showing the fraction of the total contributed by each of these counts. Figure 13.1 shows

TABLE 13.2. Same as Table 13.1 but after Doubling All the Numbers in the Lists

Leading Digit	Fraction of Occurrences	
	$1/x$	$1/x^2$
1	.302	.111
2	.176	.371
3	.125	.184
4	.098	.110
5	.080	.074
6	.066	.055
7	.058	.038
8	.052	.031
9	.044	.025

that both of these functions are largest for the lowest values of x and decrease with increasing x. Therefore it is not surprising that there are more numbers with a leading digit of 1 than with a leading digit of 2, more numbers with a leading digit of 2 than with a leading digit of 3, and so on.

I am going to take the two lists of random numbers and double every number in them. Since the original lists were of numbers between 1 and 10, the new lists will be of numbers between 2 and 20. Now I'll repeat what I did above—I'll count the number of occurrences of each of the leading digits and report these occurrences as fractions of the total.

Table 13.2 shows the results of this exercise. Compare Tables 13.1 and 13.2. For the function $1/x^2$ there is no obvious relationship between the two tables. This is what you'd intuitively expect for the results of such a seemingly arbitrary exercise. The results for the function $1/x$ on the other hand are striking—the entries are almost identical.

Without showing it here, let me state that I would have gotten the same results regardless of what number I had multiplied the lists of random numbers by. Also, had I invested the time to generate extremely large lists of numbers, I could have shown that the agreement gets better and better the lists get larger. The function $1/x$ has the unique property that the distribution of leading digits does not change (is "invariant") when the list is multiplied by any number (sometimes called a "scaling factor").

At this point I've demonstrated an interesting statistical curiosity and nothing more. Now look at Table 13.3. This table is generated by using the same procedure as above for a list of the distances to the 100 closest stars to the earth. Since this is a relatively small list, we shouldn't expect ideal agreement.[2] It does look, however, is if this list was generated using the $1/x$ distribution.

---

[2] The chapter on Statistical Sampling and Inference will discuss what accuracy you can expect from small samples of a large data set.

TABLE 13.3. Percentage of Leading Digits for List of 100 Nearest Stars

Leading Digit	Fraction of Occurrences
1	.350
2	.250
3	.100
4	.080
5	.050
6	.060
7	.040
8	.020
9	.050

TABLE 13.4. Excerpt from List of Land Areas and Populations of Countries

Area	Population	Country	Capital
250,000	31,056,997	Afghanistan	Kabul
11,100	3,581,655	Albania	Tiranë
919,590	32,930,091	Algeria	Algiers
181	71,201	Andorra	Andorra la Vella
481,351	12,127,071	Angola	Luanda
171	69,108	Antigua and Barbuda	St. John's
1,068,296	39,921,833	Argentina	Buenos Aires
11,506	2,976,372	Armenia	Yerevan
2,967,893	20,264,082	Australia	Canberra
32,382	8,192,880	Austria	Vienna
33,436	7,961,619	Azerbaijan	Baku
5,382	303,770	Bahamas	Nassau
257	698,585	Bahrain	Al-Manámah
55,598	147,365,352	Bangladesh	Dhaka
166	279,912	Barbados	Bridgetown
80,154	10,293,011	Belarus	Mensk (Minsk)
11,787	10,379,067	Belgium	Brussels
8,867	287,730	Belize	Belmopan
43,483	7,862,944	Benin	Porto-Novo
18,147	2,279,723	Bhutan	Thimphu

Is this a coincidence of just one naturally occurring set of numbers, or is there something else going on here? I downloaded a list of the land area and population of 194 countries from the Internet. Table 13.4 shows the beginning of this list, in alphabetical order by country. As an aside, one has to wonder at the accuracy of the data. For example, can the group that surveyed Bangladesh really be sure that there were at the time of the completion of their survey

TABLE 13.5. Percentage of Leadings Digits for Lists of Land Areas and Populations of Countries

Leading Digit	Fraction of Occurrences		Ideal Benford Terms
	Area	Population	
1	0.340	0.273	0.301
2	0.149	0.180	0.176
3	0.139	0.119	0.125
4	0.129	0.119	0.097
5	0.036	0.067	0.079
6	0.057	0.025	0.067
7	0.046	0.062	0.058
8	0.046	0.062	0.051
9	0.057	0.036	0.046

exactly 147,365,352 people in Bangladesh at that time? My intuition tells me that this number isn't accurate to better than about 100,000 people. In any case, since I'm interested in the leading, or most significant, digit only, this and other similar inaccuracies won't affect what I'm doing.

Table 13.5 shows the results of the leading number counts on the area and population numbers in the list (the entire list of 194 countries, not just the piece excerpted in Table 13.4). It's not obvious how well these numbers do or don't fit the Benford series, so I'd like to compare them.

Getting the ideal distribution numbers for the Benford series starting with the $1/x$ is a small calculus exercise. The resulting formula is

$$B_n = \log\left(\frac{n}{n+1}\right)$$

where B_n is the fraction of numbers whose leading digit is n. These terms are also shown in Table 13.5.

Figure 13.2 compares the ideal Benford terms to the land area and population data. Again, I've drawn continuous lines rather than showing the discrete points just for clarity. As can be seen, the fit isn't bad for such small data sets.

Can we conclude that just about any set of data will obey these statistics? No. For example, data that are normally distributed will not look obey these statistics. Uniformly distributed data will not obey these statistics. However, it has been proven that if you build a data set by randomly choosing a distribution from a list and then randomly choosing a number using that distribution, the resulting data set will obey Benford's law.

A simple caveat should be mentioned. Suppose I have a list of some measurements made up of numbers between 1.000 and 1.499; 100% of these

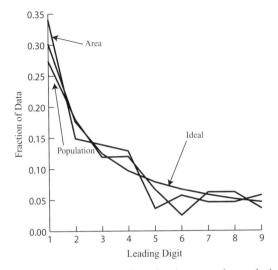

Figure 13.2. Ideal Benford fractions, land area and population data.

numbers have a leading digit of 1. If I double these numbers, 100% of the new list will have a leading digit of 2. The entire discussion falls apart if the numbers are not drawn from a list spanning multiples (at least one) of decades. Legitimate lists are numbers from 1 to 10, 1 to 100, 10 to 100, 2 to 20, 300 to 3,000,000, and so on, whereas 1 to 5, 20 to 30, and so on, are bad choices. If the list spans many decades, then the upper limit loses significance, because of the shape of the underlying $1/x$ PDF; there are relatively so few numbers coming from the large values of x that they don't contribute to the statistics in any meaningful way.

An interesting thing about data sets such as the distances to the nearest stars is that the set will obey the Benford statistics regardless of what units these distances are expressed in. This because changing of units, say from miles to kilometers or light-years, is done by multiplying every distance by a scale factor. For example, to change from meters to inches, you would multiply every number by 39.37. Since a data set obeying the Benford statistics will obey it just as well when multiplied by an arbitrary scale factor, the leading digit (Benford) statistics will not change.

A large company's collection of (say) a year's worth of business trip expense reports will obey the Benford statistics. I don't know if it's an urban legend or really the truth, but the story I heard is that the IRS has caught companies generating spurious expense statements using random number generators because they didn't realize that neither a uniformly nor a normally distributed list of numbers between, say, $10 and $1000 conforms to the correct statistical distribution.

The frequent occurrence of the Benford's Law in the world around us is really not as surprising as it might first seem. If we were to generate a sequence of numbers by repeatedly adding a constant to the previous number, for example

$$1, 1.1, 1.2, \ldots\ldots\ldots 9.9, 10.0$$

we would get a uniformly distributed sequence of numbers. A specific range, for example a range 1 unit wide, would contain the same number of members of the sequence wherever it was placed in the sequence.

This sequence, called an *arithmetic sequence*, is therefore an example of what we have been calling a uniform distribution from 1 to 10.

The sequence of numbers generated by multiplying the previous number by a constant, for example

$$1, 1.10, 1.21, \ldots\ldots\ldots$$

generates a sequence where the separation between the numbers increases as the numbers get large. This sequence, called a *geometric sequence*, is an example of a $1/x$ distribution of numbers which demonstrates Benford's Law.

Philosophically, there is no reason to believe that nature should prefer the first of these sequences to the second. It does seem reasonable, however, to assert that there is no reason for nature to have a *preferred* set of units. This would make the second sequence, whose members obey Benford's law, what we should expect to see rather than what we are surprised to see, when we look at, say, the distance to the nearest stars (whether they're measured in light-years, miles, kilometers, or furlongs).

One last word about Benford series: Figure 13.3 shows the results of adding up 10 random sequences generated using the $1/x$ PDF. As can be seen, despite the unusual properties of the $1/x$ PDF, it does not escape the central limit theorem.

PARRONDO'S PARADOX

Juan Parrondo described some games of chance where combining two losing games produces a winning game. The first of these games is a simple board game where a player moves back and forth in one dimension, with the direction of the move being determined by the throw of a pair of dice. Figure 13.4 shows the playing board for this game, and Table 13.6 shows two sets of rules for playing the game.

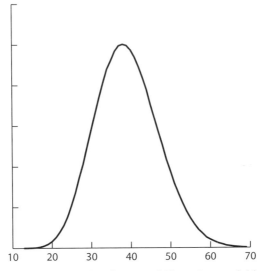

Figure 13.3. Distribution function for the sum of 10 random variables generated using the $1/x$ PDF, $1 < x < 10$.

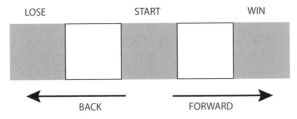

Figure 13.4. Playing board for the Parrondo's Paradox game.

Let's start using only Rule Set 1. The player starts on the center (black) square and rolls the dice. The section of Rule Set 1 labeled *black* is used. If the dice show 11, the player moves forward one box; if the dice show 2, 4, or 12, the player moves back one box. If the dice show any other number, there is no movement and the dice are rolled again. Let's assume that after however many rolls of the dice it takes, there is finally either an 11, 2, 4, or 12 and the player has moved one box forward or back.

The player is now on a white square. The section of Rule Set 1 labeled *white* is used. If the dice show 7 or 11, the player moves forward, if they show 2, 3, or 12, the player moves back. If any other number comes up, the dice are rolled again.

If the player moves forward to the box labeled WIN, then the player has won and the game is over. If the player moves back to the box labeled LOSE, then the player has lost and the game is over.

TABLE 13.6. Rule Sets for the Parrondo's Paradox Game

	Rule Set 1		Rule Set 2	
	White	Black	White	Black
Playing Rule Sets for the Parrondo's Paradox Board Game				
Forward	7, 11	11	11	7, 11
Back	2, 3, 12	2, 4, 12	2, 4, 12	2, 3, 12
Playing Rule Sets for the Parrondo's Paradox Board Game Rewritten to Show Relative Probabilities Rather than Dice Rolls				
Forward	8	2	2	8
Back	4	5	5	4

At this point it's not obvious whether the player's odds are better or worse than 1:1 in this game. In order to study the odds, I'll rewrite the rules table to show the relative probabilities rather than dice rolls. Relative probabilities are just comparisons of two or more probabilities. For example, from the roll of a pair of dice the probability of getting two is just 1/36 while the probability of getting four is 3/36. The relative probabilities of these rolls are 1 and 3, respectively. This rewrite is also shown in Table 13.6.

Looking at this latter table, there are (8)(2) = 16 ways to move forward and (5)(4) ways to move backward. The relative probabilities of winning and losing are therefore 16 and 20, so that the probability of winning is

$$\frac{16}{16+20} = \frac{16}{36} = 0.444$$

This is definitely not a good game to play.

Rule Set 2 is a bit different than Rule Set 1, but the relative probabilities turn out to be exactly the same; 16 and 20 to win and lose, respectively. This is also definitely not a good game to play.

So far I haven't shown anything remarkable. The two rule sets describe two somewhat convoluted but basically uninteresting games, both of which are losing games. Now I'll change things a bit. We'll play the game by starting with either rule set and then randomly deciding whether or not to switch to the other rule set after each roll of the dice.

The average number of forward moves is now

$$\left(\frac{8+2}{2}\right)\left(\frac{8+2}{2}\right) = 25$$

and the average number of back moves is

$$\left(\frac{4+5}{2}\right)\left(\frac{4+5}{2}\right) = 20.25$$

The probability of winning is now

$$\frac{25}{25+20.25} = \frac{25}{45.25} = 0.552$$

This is a winning game!

Another approach is to randomly switch rule sets between games. Figure 13.5 shows the probability of winning this game versus the probability of switching rule sets before each move.

The winning probability of 0.552 predicted above occurs for $p = 0.5$. This is equivalent to flipping a coin before each dice roll to determine whether or not to change the rule set. $P = 0$ is equivalent to never switching rule sets, so the probability of winning is 0.444 as predicted above, whichever rule set you happen to be using. For $p = 1$ you are switching rule sets before every roll of the dice, and the winning probability climbs all the way to about 0.58. Note that you only have to switch more than about 20% of the time in order to have a winning game.

In this board game, the operation of Rule Set 2 is dependent on the players position on the board. Another form of the Parrondo game depends on accumulated capital:

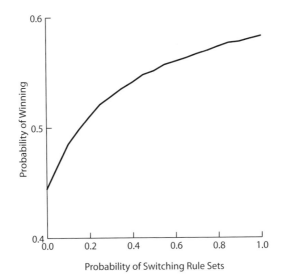

Figure 13.5. Probability of winning the Parrondo's Paradox board game versus probability of switching rule sets before every move.

Consider two coin flip games, each with their own rule sets. In both games we have the usual bet: The loser gives a dollar to the winner. The first game is very simple, it's just like the typical coin flip game except that the coin isn't a fair coin. The probability of a head is

$$p_1 = 0.5 - \varepsilon$$

where ε is a common mathematician's notation for some positive number that we can expect to be very small but still greater than zero (>0). "Very small" must of course be taken in context; a very small distance compared to the distances between stars will not mean the same thing as a very small distance compared to interatomic distances. In this case, from the context, we'll assume that ε is very small as compared to 0.5. I'll do my first example using $\varepsilon = 0.005$.

If p_1 is the probability of getting a head and I'm betting on heads, then clearly I shouldn't plan to win.

The second game is a bit more complex. There are two more unfair coins. These coins have the following probabilities of getting a head:

$$p_2 = 0.1 - \varepsilon$$

$$p_3 = 0.75 - \varepsilon$$

The choice of whether to use p_2 or p_3 is determined by how much money you've won or lost (your capital, C) when you're about to play the second game. If C is divisible by some integer that we've chosen before we start playing (call it M), then use p_2; otherwise use p_3.

I'll use $M = 3$ for my examples. This second game is a bit messier to analyze than the first game, but it can be shown that it's also a losing game.

So far I've shown two coin flip games that are only good for losing money. But what if I choose which game to play next randomly? Figure 13.6 shows the expected winnings after 100 games versus the probability (P) of picking game 1. If $P = 1$, then you're always playing game 1 and will lose money. If $P = 0$, then you're always playing game 2 and will always lose money. For a large range of P around $P = 0.5$ (randomly choosing either game), you can expect to win money. Not much money, but winning nevertheless. Figure 13.6 is based upon averaging 10,000 iterations of the 100-coin flip game.

Why does this work? Game 2 is structured so that more times than not, you're losing. However, whenever you get to a certain level of capital, you win. Game 1 is losing, but barely losing. Jumping from game 2 to game 1 after winning game 2 gives you a reasonable probability of "locking in" this gain. If the probabilities are structured correctly, you win game 2 often enough and lock in the winnings often enough, and you come out net ahead.

Can you go to Las Vegas or Atlantic City and set yourself up with some set of games that are equivalent to the above? It doesn't appear so because

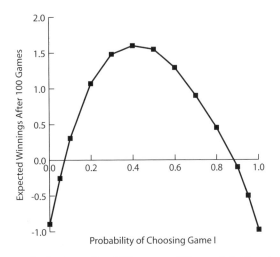

Figure 13.6. Expected winnings after 100 games of Parrondo's Paradox coin flip game versus probability of picking game I.

of the dependency of the odds of game 2 on previous results, which sometimes come from game 1. Also, you wouldn't be able to piece this together yourself from several different games because one of the games has a probability of winning of close to 75%. If I could find such a game, I wouldn't worry about Parrondo's Paradox—I'd just buy a home near the casino that's offering this game and call playing it my life's work.

On the other hand, suppose you did talk some casino owner into setting you up with this combination of games. Would they be really stupid in giving you (a set of) games that have the odds in your favor?

The standard deviations on each of the results (different values of P) are about 2. This is a very large number as compared to the expected values, and results will be quite erratic. Although the odds are truly in your favor, they're only very slightly in your favor. This is beginning to look like the "unlosable" game of Chapter 8: The mathematics says that you can't lose, but you have to start out with an awful lot of money in your pocket to cover the variations if you hope to realize the potential.

One last thought on this dice game. I can find a game in a casino where my winning probability is almost 0.5—for example, betting on one color at a roulette table. I can certainly find a game where my winning probability is almost 0.1; if I can't find one, I'm sure I could get a casino owner to create such a game for me. How about completing the picture by having a friend create the game where my winning probability is almost 0.75? I would fund this friend out of my winnings, and now I'm set to run the complete Parrondo dice game.

Unfortunately, this would be a total disaster. What happens is that I win a little bit, the owner of the "almost 0.5 probability game" wins a little less than me, the owner of the "almost 0.1 probability game" wins about 20 times more than me, and my friend, the surrogate owner of the "almost 0.75 probability game," pays for it all. Well, it looked pretty good for a couple of minutes, didn't it?

The "Parrondo Ratchet" is another example of two "losing" situations that combine to create a "winning" situation.

If some balls are placed close together on a level surface that is being shaken back and forth slightly, they will behave as what was described as random walks, aka Brownian motion. There will be an equal average of motion in all directions. If the surface is tilted upward on, say, the right side, then balls moving to the left will pick up speed (on the average) and balls moving to the right will lose speed (on the average). All in all, there will be a net motion to the left—that is, down the hill. Note that I must always stick in the term "on the average" because the shaking of the surface combined with the collisions of the balls can sometimes reverse the direction of travel of a ball for a short time.

What if I build a sawtooth structure, also known as a linear ratchet, as shown in Figure 13.7, shake it a bit and drop some balls onto it? If a ball has enough energy to climb over the top of the local hill, then it would move laterally from hill to valley to hill to valley, and so on. If I have a bunch of balls rattling around with enough total energy for a ball to occasionally climb over the top of a local hill, then I would see a gradual diffusion of the ball collection (much like the ants in Chapter 3). If the ratchet is tilted so that it's lower on the left than on the right, I would expect to see a net motion of balls to the left—that is, "down the hill."

Now let's assume that the tilted plane and the tilted ratchet are side by side and balls are getting knocked back and forth between the two as well as moving up and down the slope. Given the right parameters, it can be shown that this situation closely mimics the Parrondo coin toss. The tilted plane is

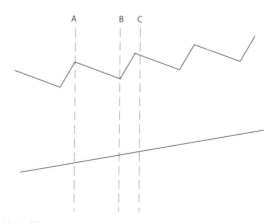

Figure 13.7. Sloped ratchet and sloped plane for Parrondo's ratchet.

game 1; the ratchet, with two different set of rules (the asymmetric teeth), is game 2. And, just as in the Parrondo coin toss, things can go contrary to what you'd expect—there can be net movement of balls up the hill.

The Parrondo ratchet system is the subject of study of people trying to learn how and if it's possible to extract energy out of a system made up of random motion. Clearly, if they're looking at Brownian motion, they're going to need some very small ratchets, but this is the age of nanomechanics, and there just might be some interesting developments coming in this field.

SIMPSON'S PARADOX

Simpson's Paradox refers to an unintuitive result that comes out of certain simple algebraic maneuvers. It is in no way as sophisticated or subtle as Parrondo's Paradox. Simpson's Paradox, however, occurs in many common situations, whereas Parrondo's Paradox, either the probability game or the coin flip version, is more commonly restricted to academic studies. Simpson's Paradox is not formally a probability topic since there need not be any randomness involved. I will, however, introduce a probabilistic interpretation to present a way of summarizing data that doesn't create the paradox.

I used an example of Simpson's Paradox in the preface, just to introduce readers to the idea that things can often be trickier than they first appear to be. I'd like to review this example, present the mathematical conditions involved, and then go on to the probabilistic interpretation of Simpson's Paradox.

Let's review the example in the preface, without the accompanying story: There are 600 women and 400 men applying to program I at a school. There are openings for 100 students in program I; 75 women and 25 men are accepted into the program. In terms of rates of acceptance, 75/600 = 12.5% of the women applicants and 25/400 = 6.25% of the men applicants are accepted into program I. The rate of acceptance of women is nearly double that of men.

There are 25 women and 75 men applying to program II at the school. Program II also has 100 openings, so all of the applicants are accepted. The rate of acceptance of women equals the rate of acceptance of men = 100%.

We added up the numbers for the two programs: 625 women applied and 100 women were accepted. The total rate of acceptance for women was 100/625 = 16%. 475 men applied and 75 men were accepted. The total rate of acceptance for men was 75/475 = 21%.

Even though the rates of acceptance for women were the same as or better than the rates of acceptance for men in both programs, the overall rate of acceptance for men was better than the overall rate of acceptance for women. In situations such as this, there is a *lurking variable* that gets lost in the summarization process. In this particular example, the lurking variable is the tremendous disparity in the number of applicants to the two programs.

Before looking at another example, let's formalize the arithmetic a bit. I'll call the two initial categories of data, the men and the women, small *abcd* and capital *ABCD*, respectively. Let

a = the number of men accepted to program I (25)

b = the number of men that applied to program I (400)

c = the number of men accepted to program II (75)

d = the number of men that applied to program II (75)

A = the number of women accepted to program I (75)

B = the number of women that applied to program I (600)

C = the number of women accepted to program II (25)

D = the number of women that applied to program II (25)

In this notation, the percentage of men accepted to program I is less than the percentage of women accepted to program I and is expressed as

$$\frac{a}{b} < \frac{A}{B}$$

The percentage of men accepted to program II is the same as the percentage of women accepted to program II and is expressed as

$$\frac{c}{d} = \frac{C}{D}$$

To our surprise, we find that the percentage of men admitted to both programs is greater than the percentage of women admitted to both programs:

$$\frac{a+c}{b+d} > \frac{A+C}{B+D}$$

Let's look at another example. In medical literature, the term *confounding variable* is used rather than the *lurking variable* seen in probability and statistics literature. In the following example, the confounding variable is the size of kidney stones.

Consider two treatments for kidney stones, I and II. For small stones, out of 87 patients treated with treatment I, 81 treatments were considered successful. Using the notation introduced above, we obtain

$$\frac{A}{B} = \frac{81}{87} = 93\%$$

Two hundred seventy patients were treated with treatment II, and 234 of these were considered successful:

$$\frac{a}{b} = \frac{234}{270} = 87\%$$

We can conclude that treatment I is more effective for small stones, that is,

$$\frac{a}{b} < \frac{A}{B}$$

For large stones, 263 patients received treatment I with 192 successes,

$$\frac{C}{D} = \frac{192}{263} = 73\%$$

and 80 patients were treated with treatment II, yielding 55 successes:

$$\frac{c}{d} = \frac{55}{80} = 69\%$$

Again, treatment I is more effective:

$$\frac{c}{d} < \frac{C}{D}$$

We seem to have a clear winner in treatments. Treatment I is more effective than treatment II for both large and small stones. However, if we try to demonstrate this overall conclusion by just adding up the numbers, we obtain $A + C = 81 + 192 = 273$ = treatment I successes out of $B + D = 87 + 263 = 350$ attempts. For treatment I overall, therefore, the success rate is

$$\frac{A+C}{B+D} = \frac{273}{350} = 78\%$$

$a + c = 234 + 55 = 289$ treatment II successes out of $b + d = 270 + 80 = 350$ attempts. For treatment II overall, therefore, the success rate is

$$\frac{a+c}{b+d} = \frac{289}{350} = 83\%$$

Again, we find that

$$\frac{a+c}{b+d} > \frac{A+C}{B+D}$$

Based on the summary, it's "clear" that treatment II is more effective.

Just out of curiosity, I cranked up my Monte Carlo simulator and created the following situation: Suppose I have the data of some study involving two groups and two treatments, or class admissions, or … I set the minimum size for each group at 50; then the maximum size for each group is a random number between 50 and a maximum size, as will be shown below. Assuming the studies are of some logical issue, I assumed that the results of the study are random numbers that are never less than 10% or more than 90% of the sample size. I then ran this scenario millions of times and counted the number of times that I got the inequalities shown above. In other words, is Simpson's Paradox such a freak situation that it takes a probability book author to search out occurrences, or is it a common enough situation that we really have to be wary of what we read?

Figure 13.8 shows the results of my Monte Carlo simulation. For low values of the maximum allowed size of each group, the probability of Simpson's Paradox occurring is low—down around 0.1%. As the value of the maximum allowed size grows, so does the probability, with values leveling out at about 1%. This is because for minimum values of 50 and a large maximum values, there are many opportunities for the size discrepancies in the groups that were seen in the examples above to occur.

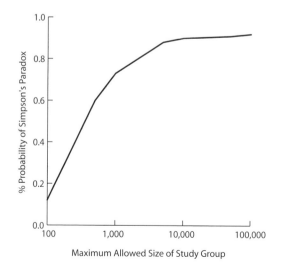

Figure 13.8. Monte Carlo simulation of likelihood of Simpson's Paradox occurring.

Now that we know that Simpson's Paradox can occur, the next logical question is, What can we do to avoid reaching these erroneous conclusions? The simplest and best guaranteed foolproof solution is to never summarize results of different studies. Unfortunately, in real life this answer isn't always available. In the case of the kidney stone treatments, for example, since we know that treatment I is more effective, we should be able to present this conclusion to the medical community without forcing busy physicians to read the details of the different studies.

The incorrect result comes about from thinking that basic weighted averages should do the trick. Going back to the first example, the fraction of men accepted to the first program is a/b, while the fraction of men accepted to the second program is c/d. We realize that we shouldn't just average these fractions because the number of (men) applicants to the first program, b, is very much greater than the number of (men) applicants to the second program. As an *improvement*, therefore, let's weight the terms by the number of men applicants:

$$\frac{\frac{a}{b}b+\frac{c}{d}d}{b+d}=\frac{a+c}{b+d}$$

which is exactly the result that got us into trouble in the first place!

Before proposing a better way, let me state that there is no one correct answer. There are several things going on at once in each of these examples, and it is necessary to specify exactly what you want to calculate before working out a formula. What I will calculate below is the ratio of the probability that one of the male applicants was accepted to one of the programs to the probability that one of the female applicants was accepted to one of the programs.

I'll define the probability that a male applicant was admitted to program I as the ratio of the number of male applicants admitted to program I to the number of male applicants to program I, with similar definitions for women and program II. Then, the ratio of the probability that a male applicant was admitted to program I to the probability that a female applicant was admitted to program I was

$$\frac{a/b}{A/B}=\frac{25/400}{75/600}=\frac{0.0625}{0.125}=0.500$$

Similarly, for program II,

$$\frac{c/d}{C/D}=\frac{75/75}{25/25}=\frac{1.0}{1.0}=1.0$$

Now, I'll weigh these by the total number of people applying to program I $(b + B)$ and the total number of people applying to program II $(d + D)$:

$$\frac{\frac{a/b}{A/B}(b+B)+\frac{c/d}{C/D}(d+D)}{b+B+d+D} = \frac{0.500(400+600)+1.0(75+25)}{400+600+75+25} = \frac{500+100}{1100} = 0.545$$

The probability of a man being admitted to one of the programs was about 55% of the probability of a woman being admitted to one of the programs; in other words, a woman had almost twice the probability of being admitted to one of the programs as did a man.

Repeating this procedure for the kidney stone treatment programs, without showing the numbers, I calculate that, overall, treatment II was 94% as effective as treatment I.

The approach above looks a little messy, but it's really very simple using a spreadsheet on a computer or even just a pocket calculator and a pad and pencil. The most important consideration of course is that it does not lead to an incorrect conclusion.

CHAPTER 14

NETWORKS, INFECTIOUS DISEASE PROPAGATION, AND CHAIN LETTERS

Some diseases propagate by your encounter with an infected person, in some cases by your breathing the air that an infected person just sneezed or coughed into, in some cases by intimate contact. In either situation the statistics of the disease propagation (and possible epidemic) is dependent upon the person–persons network; for example, you will only give AIDs to (some of) the people that you have unprotected intimate contact with. The subject of inter-people networks has been popularized in recent years by the notion of "6 degrees of separation" connecting every person to every other person on the earth. Just what this means and what the choices are in modeling these networks is an interesting subject itself and worthy of a little discussion.

The topic of "stuff" propagating along networks is a college major in itself. In addition to disease infection propagation, there is information flow along the internet, electric power flow along the national grid, and so on. On the more whimsical side, there's the chain letter.

While understanding infectious disease propagation can be literally of life and death importance whereas following the flow of dollars and letters in a chain letter's life is really more a curiosity, both of these subjects make interesting studies.

DEGREES OF SEPARATION

If I want to study how an infectious disease propagates, I need to study the statistics of how people interact. A full-detailed model of who you see on a regular basis, how many hours you spend with who, who you touch, and who you sneeze and cough at and then who these people spend time with and sneeze and cough at is an incredibly big job. Using the same reasoning that I use when studying cars in traffic and birds in the woods, however, I will try to simplify the situation to its essential components. This will let us see what's happening and let us understand the issues involved at the cost of accuracy and possibly some subtleties.

There are many ways to look at people–people interactions. The most well known of these is the "6 degrees of separation" concept. Here's how it works.

Suppose I know 25 people. Let's call this relationship 1 degree of separation. If each of these 25 people also know 25 other people, then there's 2 degrees of separation between me and $(25)(25) = 25^2 = 625$ people. If each of these (625) people knows 25 other people, then there's 3 degrees of separation between me and $(25)(25)(25) = 15,625$ people. You can see how to continue this.

If I play with the numbers a bit, I come up with $43^6 = 6.3$ billion people, which is approximately the population of the earth. In other words, if I know 43 people, and each of these 43 people knows 43 other people, and . . . , then there are no more than 6 degrees of separation between me and any other person on the earth.

The first objection everybody has to this calculation usually has something to do with a small tribe of somebodys on a lost island somewhere that doesn't interact with the rest of the world. OK, we'll agree to ignore small isolated pockets of people. The calculation is still pretty accurate.

Note that the result of this calculation is not unique. That is, if we can assume that each person knows 90 other people, then we have $90^5 = 5.9$ billion people, again approximately the population of the earth. This is no more than 5 degrees of separation. Going the other way, $25^7 = 6.1$ billion. If everybody only knows 25 other people, then there's still no more than 7 degrees of separation. Why these suppositions are more or less reasonable than assuming that, on the average, everybody knows 43 people rather than 25 or 90 people eludes me; "6 degrees of separation" is what's fixed in peoples' minds.

The biggest problem with all of this is the assertion that someone "knows 43 *other* people." This doesn't happen. Twenty years ago I would have used chain letters as an example of this assertion. Today I'm probably better off using bad jokes on the internet as an example: You receive a bad joke from someone you know (or, more likely, after the 50th such occurrence, someone you really wish you had never met). This joke is cute enough that you forward it on to a dozen or so people that you know. Within a week, more or less, you receive the same joke from other people that you know. Just where the interconnect network doubled back on itself and got to you again you'll probably never know. The point is that this happens almost every time.

I approached the simulation of people networks by taking a given number of people and randomly generating a specific number of interconnects. This let me look at some statistics of how many people know how many people, and so on. Showing a "map" of a 100,000-person network with an average of, say, 20 interconnects per person is not a sane thing to do. Instead, I'll illustrate some of the characteristics of these networks on very small networks and just quote the results for the bigger networks.

Figure 14.1 shows a 5-person network with 6 interconnects. The 5 people are shown as small circles with the person referred to by the number in the circle. The positioning of the small circles evenly on a big circle (not shown) and the consecutive numbering of the small circles is completely arbitrary. This is just a neat way to lay things out. No geometric properties such as proximity of one person to another should be inferred from the positions of the small circles on the page.

In this network, people 1, 2, and 5 have 3 interconnects (friends) each, person 4 has 2 interconnects and person 3 has only 1 interconnect. This works out to an average of 2.4 interconnects per person. Play around with this figure, or a similar figure, a bit and you'll see that there are a maximum of 10 interconnects, which works out to 2 interconnects per person. Try the same thing with 4 people and you'll get a maximum of 6 interconnects, with 6 people it's 15 interconnects, and so on. For very large networks such as the significant fraction of the number of people on the earth, we don't really care about fully interconnected networks such as this because that implies that everybody

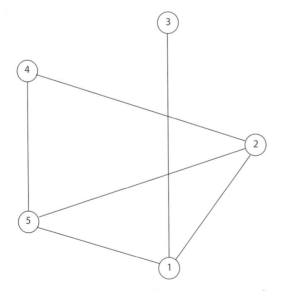

Figure 14.1. An 8-person, 6-interconnect network.

knows everybody else. If you're curious, however, the relationship between the number of people and the maximum number of interconnects is

$$\text{max interconnects} = \frac{n(n-1)}{2}$$

where n is the number of people.

An important possible property of these networks can be seen by looking at the 9-person, 11-interconnect network shown in Figure 14.2. If you inspect carefully, you'll see that although there indeed are 9 people and 11 interconnects, this really isn't a 9-person, 11-interconnect network at all. Persons 1, 2, 4, and 6 form a 4-person, 4-interconnect network and persons 3, 5, 7, 8, and 9 form a 5-person, 7-interconnect network. These two networks have nothing to do with each other.

In the real world, people networks are constructed around some guiding rules with a bit of randomness thrown in. The rule could be "members of an extended family," "fans of Elvis," Arizona Amateur Soccer League players, co-workers, or neighbors, among others. A person is usually a member of many more than one of these affinity groups. Backing off to a far enough distance and looking at the networks for such groups, however, tends to give us a picture that looks pretty random. If we're dealing with hundreds of million or even billions of people, there certainly will be isolated subnetworks that exist for no good reason whatsoever. Remember that the networks that are ultimately of interest to us are untenable to draw completely, such as the "6 degrees of freedom" network, which consists of 6 billion people and many, many interconnects.

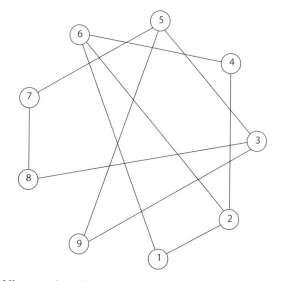

Figure 14.2. Nine people and 11 interconnects forming 2 independent networks.

PROPAGATION ALONG THE NETWORKS

Following something along the interconnect paths in our networks can have many physical interpretations and as many sets of rules. We could, for example, be interested in what happens with chain-letter money deals. In this situation you mail $1 to either everybody you know or some fixed number of people (assuming that everybody has at least this fixed number of acquaintances). In this case you'd be interested in how much money ultimately gets back to you.

Another interest would be to follow the propagation of a contagious disease. Here we need to make up some contagion rules. A simple starting point is to assume that today you see everybody you know and then tomorrow they all have the disease (and you're dead). This is pretty unrealistic, but it's an interesting extreme case with which to start studying propagation properties such as "how much time do we have to develop a vaccine or cure," and "does everybody die?" Actually, the answer to the latter question is no because of the isolated subnetworks described above.

When generating a random network for some given number of people, there are two ways to approach the randomness. First, I could fix the number of interconnects per person and then scatter the "far end" of each interconnect randomly. Second, I could fix the total number of interconnects and then scatter both ends randomly, only requiring that I don't allow duplications.

I chose the latter approach because I think it leads to more realistic results. The number of interconnects per person, by just about any set of affinity rules, is certainly not fixed. By fixing the total number of interconnects, I can "play" with this total number and study the implications of this to various consequences.

There are two sets of statistics associated with my choice of network generation algorithm. First, for a "typical" network I can look at the distribution of interconnects per person. Then I can generate a large number of networks with the same starting parameters (total number of people, total number of interconnects) and see if there really is a typical network or if each network is truly unique.

My first example is for a 10,000-person network with 50,000 interconnects. Although 50,000 seems like a large number, remember that for a 10,000-person network the maximum number of interconnects is

$$\frac{n(n-1)}{2} \approx \frac{n^2}{2} \text{ (for } n \text{ large)} = \frac{10{,}000^2}{2} = 50{,}000{,}000$$

Note that 50,000 interconnects is only 0.1% of the possible number of interconnects for this network. The first number that came out of my simulation was the average number of interconnects per person, namely, 10.0. This is more a check on the simulation than new information. If I have 10,000 people and 50,000 interconnects, remembering that each interconnect touches 2 people, then the average number of interconnects per person must be 2(50,000/10,000) = 10.0.

The next number that came out is the standard deviation of the above average, namely, 3.16. This immediately got my attention because $3.16 = \sqrt{10}$, and $\sigma = \sqrt{\mu}$ is a characteristic of a Poisson distribution.

Figure 14.3 shows the distribution of interconnects per person, superimposed on a Poisson distribution with a mean of 10.0. As you can see, the fit is excellent: The distribution of interconnects per person is indeed a Poisson distribution. Once again I have chosen to show discrete distributions as continuous lines rather than a jumble of points. These are still, however, two discrete distributions that are virtually identical.

Realizing that the distribution of interconnects per person is a Poisson distribution saves me the trouble of running the simulation thousands of times and looking at the results—I already know what the results will be.

It's not hard to imagine a town, or small city, with a population of 10,000 with each person knowing (on the average) 10 people. Now I'd like to look at how something, be it information or a disease, propagates through this town.

If I assume that each person "touches" every person on their interconnect list each day and that this touch transfers a contagious disease each day, then the disease spreading problem is identical to the "degrees of separation" problem: The statistics of the number of days it takes to infect everybody is identically the statistics of the "degrees of separation" problem.

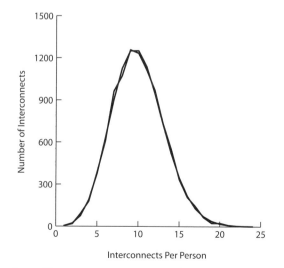

Figure 14.3. Number of interconnects per person for a 10,000-people, 5000-interconnect network.

TABLE 14.1. Infection Propagation through a Town Assuming 100% Infection Rate

Level	Number Infected	Multiplier
1	1	
2	14	13.00
3	165	10.79
4	1,588	8.62
5	7,983	4.03
6	9,997	0.25
7	10,000	

Using the same network as above (10,000 people, 50,000 interconnects) and the absolute disease spreading rules described above, I simulated the spread of disease through my little town. The results are shown in Table 14.1. The first person infected spread the disease to 13 other people. Each of these people spread the disease, on the average, to 10.79 other people, and so on. Eventually, everybody was infected.

Repeating this simulation using a network of 10,000 people and only 10,000 interconnects, I got a different result. Even though the infection is guaranteed to propagate through the network with 100% probability, the number of people infected never rose past 7941. What has happened here is that in a town where each person knows, on the average, only two other people, it's pretty likely that there are both (a) islands of interconnects that are isolated from the each other and (b) people contracting the disease and then becoming noncontagious and breaking interconnects. Repeating this same simulation over and over again gives slightly different results, but this is telling us as much about the nature of random-association networks as about disease propagation.

Returning to the network of 10,000 people and 50,000 interconnects, I next ran the simulation for a network of 20,000 people and 100,000 interconnects. The results, scaled to the number of people, were virtually identical. This is because both cases agree in the number of people, on the average, that each person knows—as a fraction of the total number of people. I repeated this latter simulation many times; and in every case at level 7, all the people were infected. Again, I am describing a characteristic of the network rather than the infection; with this density (number of interconnects per person) it is apparently highly unlikely to find any isolated subnetworks.

Now, what if it's not guaranteed that an infected person infects every person they know? This is very realistic. For one thing, you don't usually see everybody you know every day. Second, if the disease is spread by coughing or sneezing, you simply might not pass it to everybody you see. Some people will have higher levels of immunity than others and are not as easily infected. And finally, if transmitting the disease requires intimate contact, this usually translates to the disease only being passed to a small percentage of the people that, in some sense, a person knows.

Figure 14.4 shows the results of the same simulation couched in terms of the percentage of the total population (10,000 people) infected as a function of the probability of passing the infection to someone you know. Remember that this is a very simple-minded system. You get the disease on day 1, you pass it to everybody you know (with a certain probability of passing it to each person) on day 2, and on day 3 you're either dead or cured and are no longer contagious. I should point out that whether you are dead or cured is certainly very important to you but is of no consequence whatsoever to the propagation of the disease—no longer contagious is no longer contagious.

According to this figure, for probabilities of about 30% and greater, eventually 100% of the population gets infected. The simulations show that on the average it takes more days (levels of interconnect) to pass the infection to everybody when the probability is 30% than it is when it is 100%, but still eventually everybody gets infected. Below 30% probability the ultimate infection level falls, until at about 6% it drops precipitously to very few people.

Figure 14.5 shows the same results but for a more densely populated network. I doubled the average number of interconnects to 20 per person. The curve has the same general shape as that of Figure 14.4, but the numbers are shifted a bit. In particular, we need a lower probability of infection for the more densely populated network in order to see the "cutoff" of disease spread.

I could continue further with further examples, such as looking at an extremely sparsely populated network and varying the probability of infec-

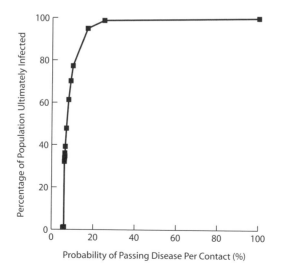

Figure 14.4. Percentage of population eventually infected versus probability of passing infection per contact (10,000 people, 5000 interconnects).

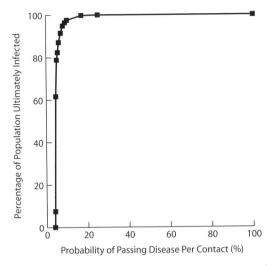

Figure 14.5. Percentage of population eventually infected versus probability of passing infection per contact (10,000 people, 10,000 interconnects).

tion. I could set up a system with two or more densely populated networks that are connected to each other by only a relatively few links—this would simulate the spread of a disease intracontinentally (the densely populated networks) and intercontinentally (the sparsely populated links between the dense networks). I could get subtle about the mechanisms and timings of infection by studying real diseases. I think, however, that the general characteristics of randomly formed networks of people has been shown.

More detailed simulations such as this have led to knowledge about real disease spread. The general conclusions drawn that you can draw from the simple simulation are valid. If a disease spreads by contact, then contact must be minimized. Tell people to stay home as much as possible, and so on. An (probably *the*) important point is the shape of Figures 14.4 and 14.5. It's not just that cutting down contagion probability will lower the rate of infection (this is pretty much common sense), the point is that when the probability gets low enough, the disease stops propagating and the epidemic dies away.

SOME OTHER USES OF NETWORKS

Diseases aren't the only thing that propagate along networks. In today's world, information transfer along networks is a major factor in all facets of communications.

The telephone system is a network of telephones and interconnects. The telephone system is of course a "managed" network in that when you want to talk to your friend the interconnection is set up. That is, your telephone line (interconnect) goes to a central office where the electrical version of your conversation is converted to digital information. This digital information then gets to "time share" on a large central interconnect to some central office near the person on the other end of your call. There are possibly several routing points that have been passed through.

The cellular telephone systems, from a network point of view, are made up of base stations that are "central offices" that connect the wired, or land-line, telephone system to radio connections to your cellular handset. The cellular system is very intelligent in that as you move in and out of range of different base stations, your call is "handed over" to the closest base station without you ever knowing that this hand-over has occurred.

The internet is an interconnection of digital networks. On the internet, your information is broken into short packets, which then travel along with other folks' messages on the "information superhighway."

In all of these information handling networks, there is always the possibility that a piece of a message will get corrupted. A lightning flash near some buried cable can, for instance, create a pulse of signal along the cable that will change some of the digital pulses. There are some very sophisticated systems for error detection and correction along these networks. Understanding these systems is an education in itself. I can, however, describe some very simple techniques and give you the flavor of what's going on.

When you speak into a telephone, the telephone handset converts your spoken words into a voltage level that varies with time as you speak (an analog signal). This voltage level travels down the wires from your telephone to the nearest central office. There, large computers process your conversation by sampling—that is, capturing the voltage in a computer's memory many times per second (the "sampling rate"). Each voltage level is then converted to a binary number (a string of 1's and 0's), and this binary number is then converted to two voltage levels which are then intertwined with the binary numbers from other conversations and sent on their way along shared interconnect lines. At the far end the reverse is done, and a voltage level versus time shows up at the handset at the other end of the conversation. The bookkeeping to keep all of this straight is very, very complicated.

One way of checking for errors, albeit a very inefficient way, is to repeat the transfer of binary information sent every, say, 0.01 second. If the repeated information is identical to the original information, then it's highly unlikely that either (or both) of the transmissions was corrupted somehow. If the two messages are different, send the message a third time. If the third message is identical to either of the first two, then the third message is most likely correct. If all three messages differ, it's time to start over again (and probably a good idea to try and find out what's going wrong).

Another method involves what's called a "check sum." A (binary digital) message of a certain length (say 0.01 second's worth of data) will have a certain number of ones and a certain number of zeros. If the message is corrupted somehow, it's unlikely that these numbers will stay the same. Therefore, after each 1 second's worth of data, send the numbers that equal the sum of the number of ones and the number of zeroes. On the receiving end, make a similar count as the data comes in. Check it against the numbers sent. If there's agreement, it's most likely that the message is accurate.

An interesting network in which things flow both forwards and backwards is the chain letter. There are many versions of the chain letter, so my description might not agree exactly with what you've seen.

The way I remember it, you get a letter in the mail. The letter has a list of six names and addresses. Your instructions are to (1) mail one dollar to each of the names and addresses on the list, (2) remove the bottom name and address from the list and add your name and address to the top of the list, and then (3) mail 6 copies of the letter to people you know, excluding any of the 6 names on your incoming list.

There are 3 ways to interpret how this will work. First, I can go with the original 6 degrees of freedom assertion that, in this situation, means that each person mails letters to 6 people who have never yet gotten this letter. Second, I can make the more reasonable assumption (as I did in the disease propagation discussion) that the first scenario never happens and you start getting repeat copies of the letter, but you throw them away. Third, I can look at what happens if each time you receive the letter you repeat the process, mailing 6 copies to 6 people that you haven't mailed the letter to before. In this latter case eventually you'll come to a stop because you just don't know an infinite number of people.

In all of these cases I have to be able to weigh in the factor that some percentage of the letter recipients will simply throw the letter away.

Figure 14.6 shows the ideal chain letter as described above ("First"). You (person X) receive a letter with names A, B, C, D, E, F. You cross out name F from the bottom of the list and add your name to the top of the list. You mail the letter to G, H, I, J, K, and L.

I'll just follow the path to person H. He crosses out E, adds H, and mails the letter to 6 people, one of whom is person M. Person M sends out the letter containing the names M, H, X, A, B, C to N, O, P, Q, R, S. And so on.

After 6 such mailings, your name is crossed off the bottom of the list and the future of this chain is of no further interest to you.

At level 1, each of the 6 people sends $1 to you, giving you $6. At level, 2 each of the $6^2 = 36$ people sends you $1, giving you $36 + $6 = $42. If everything goes perfectly, you ultimately receive

$$\$6 + \$36 + \$216 + \$1296 + \$7776 + \$46,656 = \$55,986$$

Not bad for a $6 investment (not counting postage). You have multiplied your money by a factor of 55,986/6 = 9331.

There are several things to consider here before you participate in one of these chains. I'll just worry about the probabilities involved, but there might also be legal issues—this could be considered a scam.

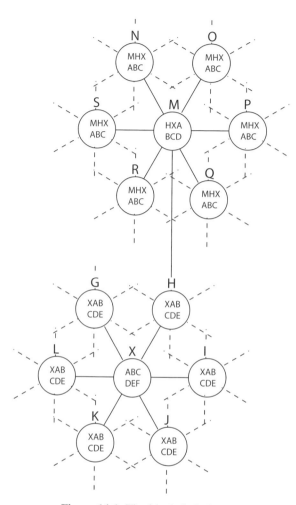

Figure 14.6. The ideal chain letter.

As was mentioned above, there have to be 55,986 people involved, only 6 of whom know you. Every one of these people has to participate fully (look out for the ones who send the letters forward but never send the money back).

In a moment I'll work through the what to expect if some people just trash the letter. After that I'll work through the cases where some of the people send one of their letters to you (using the networks developed in the previous section). Right now I'd just like to point out how, even at its most optimistic, this is madness:

At 6 levels we needed 55,986 people participating (actually, 55,987 if we count you, but let's not be finicky at this point). What if this has been going on for a while and you receive the letter at level 3—that is, 3 levels further along than is shown in the figure? Just multiply things out; for the chain to complete from your perspective, over 12 million people must participate. Three more levels and we're up past 2.6 billion people. That's almost 1/2 the population of the earth! And remember that we're still maintaining the restriction that each one of these people know only the 6 people on the list—the 6 people that they send their letters to.

So far I've demonstrated that you should never jump onto even an idealized one of these chains once a few levels have past by. Now let's keep the restriction on the people not knowing each other, but start looking at what happens when some of the people throw their letter away. In this case I'll have to talk in probabilities and standard deviations because it not only matters how many people throw their letters away, but also matters where these people are placed in the different levels. For example, let's assume that you started the chain, so we have identically the situation depicted in Figure 14.6.

Person XXX just joined the list at level 6. If he never mails you $1, you're just out a dollar. On the other hand, person A is one of your original 6 people. If she trashes your letter, then fully 1/6 the chain is broken and you're out that fraction of your winnings—that is $9331.

Figure 14.7 shows the expected value and the standard deviation of the amount of money you receive as a function of the probability of each person participating in the chain. Since not only the number of people that don't participate but also their place in the chain is important, as shown above, there is a significant standard deviation.

As would be expected (no pun intended), the expected value falls rapidly as fewer people participate. The standard deviation curve is interesting in that

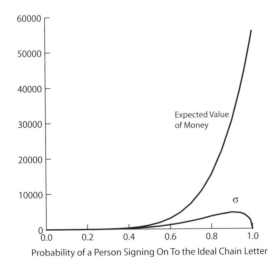

Figure 14.7. Expected and standard deviation of return for the ideal chain letter when not everybody signs on.

it must be 0 at both ends of the curve (if nobody participates, then there is 0 return with no standard deviation, whereas if everybody participates, then the return is a fixed number—again no standard deviation), so it must peak somewhere in between 0 and 100% probability. As the curve shows, it does indeed peak, at about 90% probability.

Figure 14.8 is an expanded view of Figure 14.7, showing the low probability region. This is an unusual distribution in that the standard deviation is bigger than the mean.

Figures 14.9 and 14.10 are histograms of the distribution of money collected by you for the 90% and 15% probability cases, respectively. Clearly, neither of these are normal distributions. Also, both histograms show interesting patterns, including apparently "forbidden" amounts of money. This is most apparent in the 15% probability histogram (14.10). These patterns come about because there are only a finite number of paths through the 6 levels of the chain letter, and only certain total amounts of money received can occur. Rather than showing this in its full complexity (because I don't think that knowing the actual amounts is of much interest, whereas understanding why this happens is), let me just look at a 2-level chain letter.

In a 2-level chain letter, if everybody participates, then I receive $6 + 6(6$) = $6 + $36 = 42. If everybody doesn't participate, then I can have only certain combinations, such as 1 person from level 1, 1 person from level 2, 3 people from level 1, 2 people from level 2, and so on.

Looking at the first of these, if 1 person from level 1 participates, I get $1 from this person. Level 2 then consists of only 6 letters received rather than

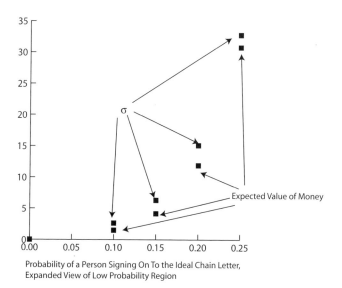

Figure 14.8. Expanded view of the low probability region of Figure 14.7.

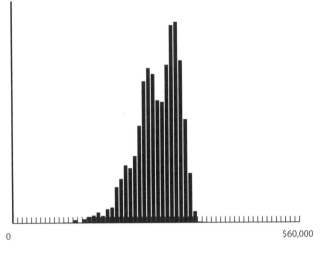

Money Collected For 90% Participation Probability

Figure 14.9. Histogram of money collected for 90% participation probability.

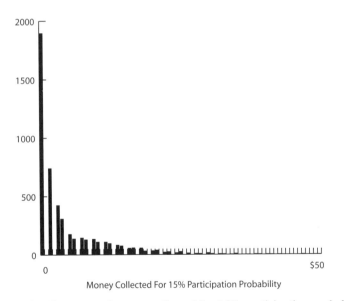

Money Collected For 15% Participation Probability

Figure 14.10. Histogram of money collected for 15% participation probability.

the full 36 possible, and if only 1 of these people participates, then I get $1 from this person, giving me a total of $2.

All of the analysis thus far seems to show that while getting the full $55,986 from the chain letter is unlikely, getting a pretty sizeable amount of money isn't difficult. For a 50% participation, the expected amount received is about $1100, which isn't bad for an hour's work addressing envelopes. The curve falls quickly, however, and at 20% participation the amount falls to about $12. This is beginning to look like a waste of time. On top of all of this, we are still dealing with the totally ridiculous assertion that we could find over 55,000 people, all of whom would mail letters only to people who would only receive this one letter—especially when each person only sees the 6 names on the list they receive, so they couldn't even try to avoid duplication.

The more realistic chain letter—that is, the chain where we acknowledge that people tend to know each other in groups rather than in random links—will behave pretty much like the chain letter above with a random probability of someone trashing the letter. If there's a random chance that the chain will work its way back to you and you trash the letter, then this isn't different from the chain reaching a new person who trashes the letter.

NEIGHBORHOOD CHAINS

The last topic I'd like to look at in this chapter is networks with not-quite-random interconnects. That is, I'll look at networks where it's more likely that you know someone who lives near you than someone who lives all the way on the other side of town. In order to do this, I must assign an "address" to every person. I'll do this by putting my map on a huge piece of graph paper—that is, on an X–Y grid—and assign a unique point, that is unique values of X and Y to each person.

You could, for example, substitute "is related to you" for "lives near you" and a family tree for address distances without overall patterns changing very much. In other words, using physical distance is really a good example of many other types of affinity relations. What happens when you mix physical distance with affinity relations that spread all over town, with membership in the community choir, and so on, is that you start forming networks that simply look like the random ones discussed earlier in this chapter.

In the real world we have neighborhoods building up towns, a bunch of towns with low population density between them building up a region, and so on. In order to just show what's going on, however, I'm just going to look at a region with 225 people. This is a ridiculously small number of people for simulating real situations, but it's as big a population as I can show graphically on a single sheet of paper with useable resolution.

Figure 14.11 shows my town. Each "X" represents the home address of a person. I placed all of these homes on a simple square grid, which again is not realistic. I chose to do this because it lets me make the points that I'd like to make without blurring the picture with patterns of homes. Figure 14.12 shows this same town with 200 randomly placed interconnects. This works out to each person knowing, on the average, 400/225 ~ 2 other people. This is a

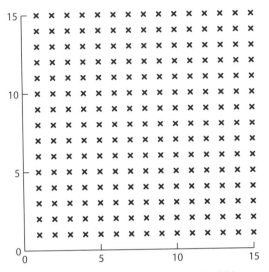

Figure 14.11. *X–Y* grid showing the locations of 225 homes in town.

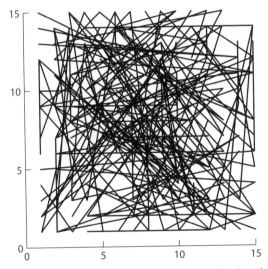

Figure 14.12. *X–Y* grid of 225 homes with 200 randomly placed interconnects.

ridiculously small number of interconnects per person; but if I used a larger number, the graph would be so matted that there would be nothing so see.

If you look in the lower right corner of this figure, you'll see a person who knows only one other person. If you look in the upper right corner, you'll see (or rather you'll notice because there's nothing to see) a person who doesn't know anybody. These occurrences are very unlikely when there is a more

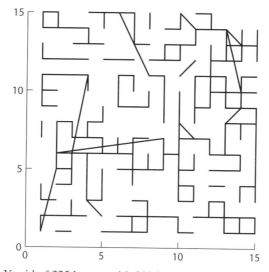

Figure 14.13. *X-Y* grid of 225 homes with 200 interconnects favoring near-neighbors.

realistic (i.e., higher) number of interconnects per person, but not at all unlikely in this situation.

In Figure 14.13, I show what happens when I constrain the random selection of interconnects. I know that the farthest possible distance between homes is upper left to lower right (or equivalently, upper right to lower left since I laid out a square). I know that the closest that two homes can be is next to each other on the grid. I can therefore look at any two homes and come up with a relative distance between them (relative to the smallest and largest possible distances). I modified the random selection of interconnects to include a scaling that (a) forces a high probability of creating an interconnect if the separation between the two homes is small and (b) forces a low probability if it's large. There are 200 interconnects, just as in the previous figure. These two figures, however, look very different.

In Figure 14.13 there are not only small isolated neighborhoods, but also several cases of neighborhoods with only 1 interconnect between them. To the extent that this approximates some real situations, it's easy to see how it's not that easy for contagious diseases to get to everybody. If the person on either end of the interconnect between otherwise isolated neighborhoods is away that week, or recovers (or dies) from the disease before he gets to visit his friend at the other end of the interconnect, the chain is broken.

I already mentioned the value of studying these types of networks with regard to understanding disease propagation. Working with telephone, or data, or power network simulations is also very useful because it lets designers study network capacity and failure probabilities. Since resources are always limited, it's important to minimize the probability that a local event such as an earthquake or a fire can shut down an entire network because the remaining branches either don't make the necessary connections or are themselves overloaded and fail. This has happened several times over the past 20 years or so to pieces of the U.S. power grid, and it's not pretty. The internet, on the other hand, has significant rerouting capabilities. Often, totally unbeknownst to us, different pieces of the web page heading our way are actually traveling along very different routes—sometimes even arriving out of time sequence (fortunately the system reorganizes them for us).

CHAPTER 15

BIRD COUNTING

Last summer my wife and I were taking part in a nature walk near a preserva-
tion area not far from our home. We came upon a sign that read something
like "Nesting Area of the Cross-Eyed Beige Fence Sitter Bird." This is an
endangered species. There are only 1327 of these birds alive." While the rest
of our group lamented the possible loss of this species, my mind started down
the all-too-familiar dangerous path of wondering about the data: "How on
earth do they know that number to anything near four significant figures?"
The best answer I could get from our naturalist–guide (whom I suspect was
making up the answer on the spot) was "We count them."

Before proceeding, I must confess that I know almost nothing about birds
or their habits. My daughter had a cockatiel when she was a child, but I doubt
that this does much to qualify me as a bird expert. I will use some assumed,
but hopefully not unreasonable, characteristics of birds living in a wooded area
as an exercise in examining what we really know about the quantity of something that we can't count in a well-behaved manner.

A WALK IN THE WOODS

A leisurely walk along a path in the woods should take us about four miles in
two hours. For convenience, I'll make the path straight. This shouldn't affect
the results because the birds' nests and flight paths will be randomly gener-

Probably Not: Future Prediction Using Probability and Statistical Inference,
by Lawrence N. Dworsky.
Copyright © 2008 John Wiley & Sons, Inc.

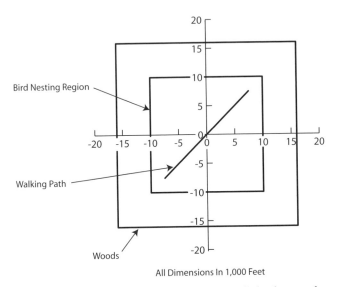

All Dimensions In 1,000 Feet

Figure 15.1. Layout of the bird counting walk in the woods.

ated; meandering the walking path would just move some of the relative distances (bird to person) around without affecting the statistics.

Assume that the species of birds we want to count are nesting in these woods. The number of birds will be a variable that we'll study. The locations of the birds' nests will be a random variable in a Monte Carlo simulation. Figure 15.1 shows the region under scrutiny, the path of our walk, and the region where I'll be (randomly) placing birds' nests.

A problem with bird counting is that if you can't catch and tag a bird, how do you know you haven't seen the same bird twice, or thrice, or … ? Also, if you have just lost sight of the bird that you have been watching because of some tree branches, and then a couple of seconds later you spot a bird that looks the same, in the same general location, how do you know if it's the same bird or a different bird? This is where our trusty tool, the Monte Carlo simulation, comes in. If we have an idea of bird flying habits and an idea of how likely it is that we spot a bird (and then lose sight of the bird) depending on its distance from us, then we can simulate the situation and get an idea of how many birds there are in an area versus how many birds we count when walking through the area. We also get a confidence interval—that is, how well we really know what we think we know.

A MODEL OF BIRD FLYING HABITS

A quick study of the subject (I checked a couple of sites on the internet) taught me that many, if not all, small birds fly in a straight line, point-to-point. When a bird alights on a branch or telephone wire or whatever, he may take off almost immediately or sit for a while. My birds (that is, the birds of this simulation) sit for some period of time between 2 and 20 seconds, with the actual time being a random variable uniformly distributed in this period.

My birds all fly at 15 miles per hour. They fly from one branch to another, flying for up to 60 seconds. The actual flight time is a uniformly distributed random variable.

All of the birds start at their nests. They leave flying in a random direction. After each flight they "note" how far they are from their nest. Their new direction is not quite randomly chosen; it's weighted by how far they are from their nests, with 5000 feet (essentially a mile) being as far as they will allow themselves to go. The farther they are from their nests, the more likely they are to head in a direction that takes them closer to, rather than further from, their nests. If a bird gets close to a mile away from its nest, its next flight will be aimed directly at its nest.

A little trigonometry and algebra are required to understand how this part random, part deterministic flight direction was calculated. If you're not interested in the details of the math, just skip right over to the next section.

Consider Figure 15.2. I've put the bird's nest at $X = 0$, $Y = 0$ (the "origin") on a graph. The actual location for the purposes of this discussion doesn't matter; I'm only looking at the relative position of the bird with respect to its nest. The standard way of measuring an angle is to start at the $+X$ axis and rotate counterclockwise until you intercept a line going from the origin to the point in question. Using this notation, a bird starting at the origin and flying

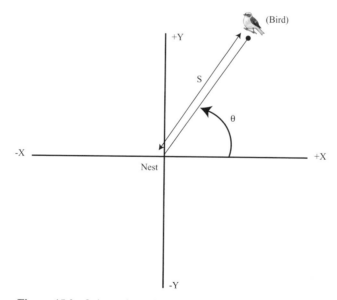

Figure 15.2. Orientation of the bird with respect to its nest.

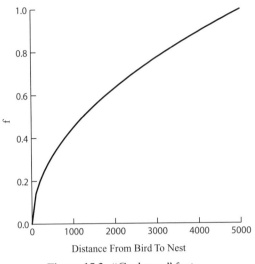

Figure 15.3. "Go-home" factor.

in the $+Y$ direction is flying at a 90-degree angle. A bird starting at the origin and flying in the $-X$ direction is flying at a 180-degree angle, and so on. Note that this system has some ambiguities associated with it because the circle repeats every 360 degrees, -90 degrees is the same as $+270$ degrees, $+180$ degrees is the same as -180 degrees, and so on. This ultimately doesn't cause any problems, you just have to be a little careful with your bookkeeping.

In Figure 15.2, the bird is at an arbitrary angle, which I'll call θ.[1]

If, starting from its current position, the bird were to fly off randomly, we could get a random direction by choosing the angle as a random number uniformly distributed between 0 and 360 degrees. If the bird were to fly from its current position directly toward its nest, it would fly at an angle of $\theta + 180$ degrees.

I will define a "go-home factor," f, as

$$f = \sqrt{\frac{s}{5000}}, \qquad 0 \le s \le 5000$$

where s is the distance from the bird to its nest, in feet.

Figure 15.3 shows this "go-home" factor. The form of the function—that is, the square root—is just an intuitive choice of mine. As you can see in the figure, when s is small (compared to 5000), f is very small. When s gets close to 5000, f gets close to 1.

---

[1] For some reason it's conventional to denote angles by Greek letters, with θ and φ being very popular choices.

If R is a random number whose distribution is uniformly distributed between 0 and 360, then the bird will take off from its perch and fly at an angle φ given by

$$\varphi = (\theta + 360)f + R(1 - f) = (\theta + 180)\sqrt{\frac{s}{5000}} + R\left(1 - \sqrt{\frac{s}{5000}}\right)$$

Let's look at how this angle is calculated. When $f = 0$, we get $1 - f = 1$, and then we have $\varphi = R$. This satisfies the requirement that when $f = 0$ (which happens when $s = 0$), the bird flies at a random angle between 0 and 360 degrees. When $f = 0$, we obtain $1 - f = 0$, and then $\varphi = \theta + 180$. This happens when $s = 5000$ and satisfies the requirement that when the bird gets to 5000 feet away from it nest, its next flight is directly toward its nest.

For intermediate values (i.e., $0 < s < 5000$), from Figure 15.3 we see that $0 < f < 1$. In this range, as s gets larger, the bird is more and more likely to fly toward home.

As an example, suppose that $\theta = 90$ degrees. The flight home would be at the angle $90 - 180 = -90$ degrees ($= 270$ degrees). Figure 15.4 shows the PDFs of the angle that the bird flies, φ, when the bird takes off at several different distances from home, for several values of s. As may be seen, the PDF starts with a distribution of all angles (0 to 360 degrees) when $s = 0$ and then narrows about 270 degrees as s approaches 5000 ft.

Figure 15.5a shows the first few legs of a bird's flights for the day. Figure 15.5b shows the bird's distance from its starting point versus time. Here you can see the wait times at the various perching points. From these figures, it's hard to tell that this isn't a simple Brownian motion situation—it appears that

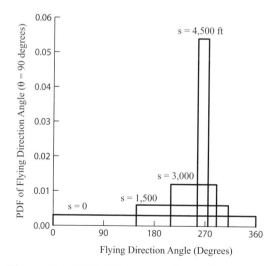

Figure 15.4. PDF of flying direction angle versus S.

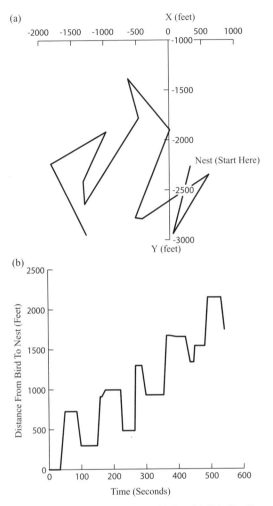

Figure 15.5. (a) Map of first few legs of bird flight. (b) Bird's distance from starting point versus time.

the distance from the starting point tends to be growing with time. This is correct because so long as the bird is much less than a mile from its nest, its flight directions are just about random.

Figure 15.6 shows 2 hours of bird flight. In this figure it is clear that our bird is not demonstrating Brownian motion; it wanders more or less randomly but really doesn't like to get too far from its nest.

Also shown in Figure 15.6 is a piece of our walk through the woods. This figure shows that both we and our bird pass through the same region at some time. What the figure doesn't show is whether or not the bird ever gets close enough for us to spot—it might be somewhere else when we're coming through.

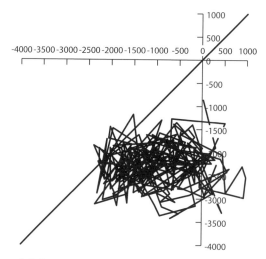

Figure 15.6. Map of 2 hours of bird flight and portion of walking path in same region.

SPOTTING A BIRD

In order to see if and when I'm close to our bird, I must calculate the distance between me and the bird as a function of time. Figure 15.7 shows the result of this tabulation when the distance is less than 1000 feet. At about 2400 seconds (40 minutes) into our walk we get less than a few hundred feet from the bird several times. In one instance we practically bump into it.

Now I must consider what it means to spot the bird. We're in a wooded (although not terribly heavily wooded) area, so this is not obvious. I'll start by saying that if the bird is very close, it's a sure thing that we'll spot it. I created the "spotting" function shown in Figure 15.8 to reflect this situation. I assumed that when the bird gets within a few hundred feet of us, we'll spot it. When it's farther away, the probability of us spotting it gets lower (overall distance issues as well as blocking trees, etc.) until at about 800 feet the probability of spotting the bird is nil.

Once the bird is spotted, I need to consider when we'll lose sight of it. This is the "losing" function, also shown in Figure 15.8. Once we see the bird, it's likely that we'll be able to keep our eyes on it for a further distance than the distance at which we first spotted it. This is reflected in the figure in that the losing probability is close to 0 for small distances and doesn't start climbing until about 600 feet. It then climbs rapidly until at about 900 feet it's pretty certain that we'll lose sight of the bird.

I added a couple of other caveats to the simulation. First, if the computer predicts that a bird was in sight for less than 5 seconds, I don't count it as a sighting. This seems to describe the fact that we need a few seconds with the bird in sight to be sure that we're really looking at bird (on a branch?) and

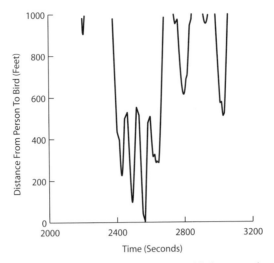

Figure 15.7. Distance from person to bird versus time.

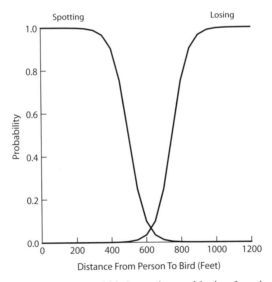

Figure 15.8. Assumed bird spotting and losing functions.

that it's the right kind of bird. Second, once we've lost sight of a bird, I incorporate a 5-second "dead" period when we're not looking for birds. This is intended to reflect the fact that if we've just lost sight of a bird and we almost immediately see another bird, it's really probably the same bird again.

Putting all of this together, the simulation predicts that we'll spot this bird five times. I don't know how we'd decide whether we saw one bird five times, five different birds, or something in between. Before concluding that we're horribly overcounting the number of birds in the woods, however, I have to admit that I "rigged" this run of the simulation so that our walking path would intersect the bird's flight area. Looking back at Figure 15.6 and noting that this figure only shows a section of the region we're considering, it's actually unlikely that we even encounter a bird.

PUTTING IT ALL TOGETHER

The way to get the real picture of what all these assumptions lead to is to move from a 1-bird simulation to the full Monte Carlo simulation. That is, I'll repeat the same walk over and over again, each time randomly placing the bird's nest somewhere in the region. This will let me look at the statistics as a function of how densely populated this region is with these birds. Figure 15.9 shows the first five bird flight patterns of this full simulation. Note that two of the trajectories overlap a bit, and there really are five trajectories shown. Looking at this figure, we can guess that two of the birds will probably get spotted at least once each, one of the birds might get spotted, and two of the birds will probably never get spotted.

Going forward with the many bird simulation, there is the caveat that I am never considering an extremely high density of birds. For example, when you're right under a flock of several hundred migrating birds settling into trees for the night, it's very hard to count birds—there's just an amazing amount of activity and simply too many birds moving around. In this case I would suggest taking a few photos—that is, "freezing" the bird movement and then learning to extrapolate from studies of these photos.

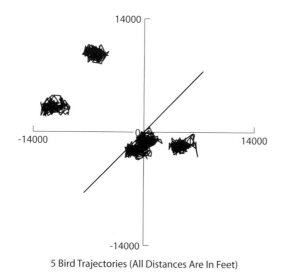

5 Bird Trajectories (All Distances Are In Feet)

Figure 15.9. Bird flight paths for the first 5 birds of the Monte Carlo simulation.

TABLE 15.1. Results of Bird Spotting Simulation for Various Numbers of Birds

Number of Birds in Region	Average Number of Sightings	Standard Deviation About Average	95% Confidence from Data	95% Confidence $\mu + 2\sigma$
1	.382	1.25	(0 to) 3	2.88
5	1.91	2.75	(0 to) 7	7.40
10	3.82	3.86	(0 to) 11	11.5
15	5.73	4.75	(0 to) 15	15.2
20	7.64	5.50	(0 to) 18	18.6

The Monte Carlo simulation was run 100,000 times. Since each bird is independent of the other birds, the average number of sightings and the standard deviation of the data about this average for different size bird populations (number of birds flying in the simulation) is easily obtained from 100,000 runs of 1 bird flying as was discussed above. It is only necessary to scale the average by the number of birds flying and the standard deviation by the square root of the number of birds flying. This assertion was verified by actually running these different situations. The results are tabulated in Table 15.1.

These results are of course intimately related to all the assumptions made: how fast the birds fly, how easy or hard it is to spot a bird, and so on. This means that the actual numbers coming out of the simulation cannot be quoted as "universal wisdom." On the other hand, the simulation does illuminate the nature of the issues involved.

For one bird flying, the expected number of sightings is a little less than 0.4, with a standard deviation on this expectation of 1.25. The 95% confidence factor is from 0 to 3 birds. The rest of the numbers follow as is shown in the table. Figure 15.10 is a histogram of the PDF of sightings when there are 15 birds flying in the region. The expected number of sightings is about 6; but with a standard deviation of 4.75, the 95% confidence interval is about 15. In other words, if you see 6 birds, your confidence interval as to how many birds are out there is 0 to 15. In this case, "out there" means that if you were to take the same walk through the woods tomorrow, you would expect to see somewhere between 0 and 15 birds. Just how many birds there really are involves scaling your results by the size of the forest. As it is, you couldn't really see all the birds that were there because they never flew far enough from their nests to intercept your walking path.

The simulation thus far is interesting, but it sort of runs backward as compared to how things work "in the field." I've shown interesting graphs of what I should expect to see if there are 2, or 14, or whatever number of birds in the woods. What I need to do is to look at the data "backwards." That is, what I'd like to see are the probabilities of how many birds are out there as a function of how many birds I see on my walk.

Fortunately, this is an easy job to do—on a computer. Since my simulation generated 100,000 pieces of data, no one would want to do this by hand. It's

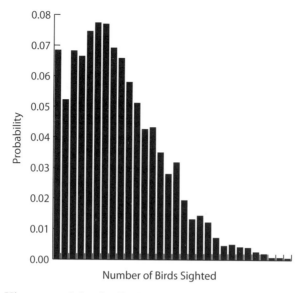

Figure 15.10. Histogram of the distribution of number of sightings when there are 15 birds in the woods.

strictly a bookkeeping job, so I won't go through the details of how to reorganize the existing data.

Figure 15.11 shows the probabilities of there being various numbers of birds in the woods if I don't see any birds on my walk. This is of course discrete data; sighting 3.2 birds or there being 4.7 birds out there doesn't mean anything.

This figure cannot be a PDF. If there are no birds in the woods, then the probability of not sighting any birds on a walk through the woods has to be 1 (100%). Once one of the points on a discrete graph has a value of 1, however, all the other points must have a value of 0 if all the probabilities are to add up to 1.

Figure 15.11 passes common-sense tests for how it "should" look. First we have the observation above: When there are no birds in the woods, it's absolutely certain that there won't be any legitimate bird sightings. [One factor I did not include in the simulation is a probability of false sightings—a "bird-like" tree branch stub or another species of bird, or a flying squirrel.] Then, the more birds there are flying around, the less likely it is that there will be 0 sightings. If there's only one bird flying around in an area that's pretty large as compared to my field of view from my walking path, then it's pretty unlikely that I'll see it; if there are 20 birds flying around, it's pretty unlikely that I won't see at least 1 of them, and so on.

Figure 15.12 shows the probabilities of different number of birds when 3 birds are sighted on the walk. The probability of there being 0 birds in the

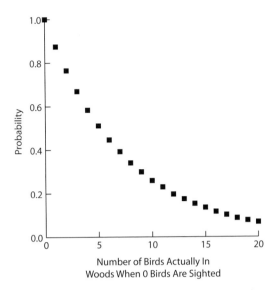

Figure 15.11. Probability of number of bird sightings versus number of birds in woods when no birds are sighted on walk.

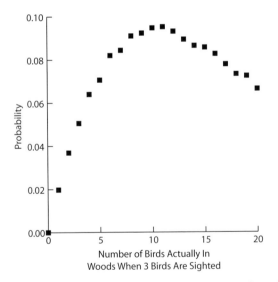

Figure 15.12. Probability of number of bird sightings versus number of birds in woods when 3 birds are sighted on walk.

woods is also known exactly in this situation; if 3 birds are sighted, then we know that it is absolutely impossible for there to be 0 birds in the woods. This probability graph peaks at about 11 birds. I can't do a confidence interval of calculation (except for a very narrow, high confidence interval of about 11

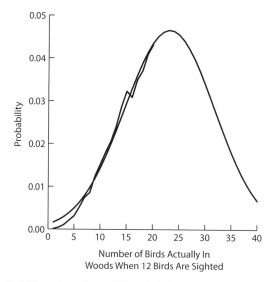

Figure 15.13. Probability of number of bird sightings versus number of birds in woods when 12 birds are sighted on walk.

birds) because my data only go up to 20 birds and it's pretty clear that a 95% confidence interval includes some of these data.

Figure 15.13 shows the probabilities of different number of birds when 12 birds are sighted on the walk. In this case the numbers are getting large enough that a normal distribution is probably a good match to the data. A best-fit normal distribution is shown superimposed on the data. I am showing continuous curves rather than points because between the two sets of data there are a lot of points and the graph gets too busy to see clearly.

Once I have a best-fit normal curve to the data, I can easily find the expected number of birds and my confidence interval. The expected number of birds is approximately 23 and the standard deviation is approximately 8.5. My confidence interval is therefore about 23 ± 2 (8.5), or from 6 to 40.

The above calculation points out the difficulty in this type of measurement very well. The 95% confidence interval is 34 (birds) wide. The expected number of bird sightings is only 23. When the confidence interval is wider than the expected value, things get "iffy." 40 is close to 6 times 6. This is a very large range of uncertainty.

This chapter is described as a simulation of bird sighting counting. While the simulation is couched in terms of a walk through the walks counting a specific type of bird, the intent is to point out some much more general things. First, when talking about physical measurements, there is a relationship between the number of significant figures quoted and the resolution of the measurement. If I quote 1327 birds, then I'm implying that *I know that it's not* 1326 or 1328 birds. If my confidence interval is plus or minus about 17 birds, then I really don't know this number to 4 significant figures.

The correct way to write this would be to specify the confidence interval. Alternatively, if I were to write 1.33×10^3, then I'm specifically quoting only 3 significant figures of resolution. This notation is not popular among the general public, so a looser but more accessible way to say it would simply be "about 1330." The 0 is actually specifying a fourth significant figure, but the "about" tends to get the message across.

As a human activity, this walk through the woods and the bird counting during the walk is not a mentally challenging activity. If anything, it's very straightforward and simple. However, look at the number of assumptions I had to come up with before I could correlate how many birds we saw in the woods to how many birds are actually there. If I wanted to push this analysis further, I could study how sensitive the results are to each of these assumptions; if I got one or more of them a little bit wrong, my results might be total nonsense. Also, I didn't factor in the weather or the time of day, among other things that might affect these birds' flying habits. The confidence interval of how many birds are actually there is probably quite a bit larger than I calculated.

The point here is that coming to conclusions based on studies of complex situations, especially characteristics of living creatures, usually involves many variables, each variable having its own mean and standard deviation. The standard deviation of each variable contributes to the standard deviation of your results. If you haven't characterized all these variables, then you are underestimating the standard deviation of your results. You probably don't know as much about what you just studied as you think you do.

CHAPTER 16

STATISTICAL MECHANICS AND HEAT

The air around us is made up of many, many molecules of gas, all moving around randomly (Brownian motion). These molecules collide, and their collisions can be pictured as the collisions of hard spheres (e.g., bowling balls). Freshman physics students learn to calculate the effects of the collisions of two spheres on the trajectories and speeds of the spheres, but what can be calculated when there are billions of these spheres careening around in every cubic centimeter of air? The answer to this question lies in the fact that while individual gas molecules each behave quite erratically, a room full of gas molecules behaves quite predictably because when we look at the statistics of the molecules' activity, we find that the erratic behavior tends to average out. This is another way of saying that the standard deviation of, say, the x-directed velocities as a fraction of the number of molecules involved is very, very small. The field of physics that starts with looking at the behavior of individual gas molecules and then predicts the behavior of gases is called *Statistical Mechanics* in recognition of the fact that it is concerned with building up mechanical models from the statistics of the behavior of billions of gas molecules while not being unduly concerned with detailed knowledge of the behavior of any single one of these molecules. The concepts of gas heat and temperature, among others, arise from this study and lead to the study of *thermodynamics*, which is concerned with the movement of heat and how we can (or sometimes cannot) get this movement of heat to do some useful work for us.

By introducing just a few principles of mechanics and using some drastic simplifications, I can derive a very good picture of the basics of gas properties and the fundamentals of thermodynamics.

STATISTICAL MECHANICS

The first thing I need to do is to introduce the concepts of the momentum and kinetic energy of a gas molecule. Both momentum and kinetic energy are related to the mass of the molecule and its velocity. I'll restrict the discussion to two-dimensional situations. Since we live in a three-dimensional universe, this is obviously a simplification, but it allows me to describe motion by drawings on a page of paper and the concepts being discussed aren't really compromised.

Velocity is a measure of how fast something is moving along with the direction of this motion. A physical quantity with both a magnitude and a direction is called a *vector*. Figure 16.1 shows a vector **V**. The length of the line represents its magnitude. If **V** is a vector representing velocity, then the length of the vector represents the speed of object being studied. The direction of the vector, from the tail to the (arrow) head, represents the direction of the velocity (which way the object is going) with respect to an XY graph, or coordinate system. Since the moving object is moving with respect to both the X direction and the Y direction, we say that **V** has components in both the X and Y directions. These components are shown as V_x and V_y, respectively.

Newton's famous laws of motion say that if no external force is applied to a moving body, its momentum, which is just the product of the mass and

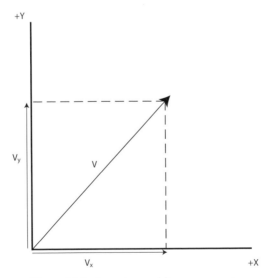

Figure 16.1. A vector and its components.

the velocity ($m\mathbf{V}$), never changes. You've probably heard this as "A body at rest tends to stay at rest, a body moving in a straight line tends to keep moving in that same straight line." Note that since momentum is just velocity multiplied by a number (a "scalar" quantity), momentum is also a vector quantity. This means that Newton's first law applies to both the X and the Y components of the momentum vector. Another way of stating this same law is that in the absence of external force(s), a body's momentum is *conserved*. Saying that a vector quantity is conserved implies that each component of the vector (X and Y here, X, Y, and Z in general) is individually conserved.[1]

Newton then went on to say that if you apply a force to a body (and force is also a vector quantity), then the force will equal the rate of change of the momentum. Since the mass isn't changing, the velocity must be undergoing a rate of change, which is called acceleration. You've probably seen this as $F = ma$.

Another property of a moving body is that since we (or somebody) put some work into getting it moving, it has stored energy known as its *kinetic energy*. Without going through the derivation, the kinetic energy of a moving body, usually referred to as T, is

$$T = \frac{1}{2}mv^2$$

The v in the above equation is the magnitude, or length, of the velocity vector—that is, the speed. Since the square of a real number is never negative, kinetic energy can be zero or positive, but never negative. Energy, kinetic or otherwise, is a scalar quantity (a fancy word for a simple number).

A basic principle of Newton's picture of our physical universe is that energy is conserved. You can put it into a body, you can get it out of a body and put it somewhere else, but you can neither create it nor destroy it.

Now let's look at what happens when two gas molecules collide. Just to keep things simple, I'll assume that both molecules are of the same type of gas, so that they have the same mass. When two bodies collide, the momentum is always conserved. That is, the total momentum of the two bodies before the collision is the same as the total momentum of the two bodies after the collision. The collision of two gas molecules is considered to be an *elastic collision*. This means that in addition to the momentum being conserved, the total kinetic energy is conserved. The combination of these two conservation rules

[1] This is obvious when you realize that while a body's velocity is unique, the coordinate system it's referred to is arbitrary. In other words, I could always line up my coordinate system so that the body is moving in the X direction. In this case the X component of the momentum is identically the total momentum. The individual vector components must therefore be conserved, regardless of the orientation of the coordinate system. Another thing I should point out here is that the mathematician's definition of "obvious" is *not* that you realize the conclusion almost immediately, but rather that after pondering it for a while, you realize that it simply could not have come out any other way.

lets us calculate which way the masses will move after a collision based on their velocities and how they collide.

Several years ago, there was a popular desk toy consisting of a wooden frame from which a dozen metal balls were suspended by strings, in a line. The behavior of these balls exemplifies elastic collision rules for small spheres (which is how we're picturing gas molecules).

Figure 16.2 shows the simple situation of two spheres coming directly at each other, at the same speed, from opposite directions. Before the collision, one sphere was moving to the right, the other to the left, at the same speed (1 foot/second in the figure).

There is no Y-directed momentum because there is no Y motion. If we call velocity in the $+X$ direction positive, then velocity in the $-X$ direction is negative. If the mass of each sphere $= 1$, then the X momentum is $+1 - 1 = 0$. In other words, this two-sphere system has no momentum, even though it definitely has energy. After the collision, we see the same thing: There is neither X nor Y momentum.

Since kinetic energies are never negative, they simply add. The total kinetic energy of the 2-sphere system is the same after and before the collision. Due

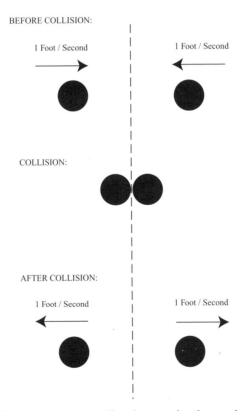

Figure 16.2. Two spheres moving directly at each other at the same speed.

to the symmetry of this example, the individual kinetic energies of the two spheres is also the same after and before the collision. This will not always be the case, and I will show that the transfer of (some of the) individual energies while keeping the total constant will be responsible for the basic properties of gases.

Figure 16.3 shows another simple example that is typically a surprise to an observer (the surprise only happens the first time the observer sees it). In this example, a moving sphere collides directly with a stationary sphere. The moving sphere comes to a complete stop and the stationary sphere moves away, at the same velocity as that of the moving sphere before the collision. Momentum is conserved because it had been +1 (the moving sphere) and 0 (the stationary sphere) and it becomes 0 and +1. Similarly, there is a complete transfer of energy with the total energy remaining constant.

Figure 16.4 shows the general two-dimensional case. Two spheres are moving at two arbitrary velocities, and they collide at a glancing angle. Calculating the resulting trajectories of the two spheres is not difficult, but it is difficult to summarize the results of the calculations. Total momentum and

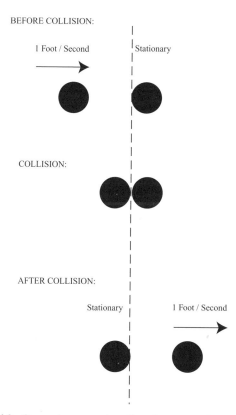

Figure 16.3. One sphere moving directly at a stationary sphere.

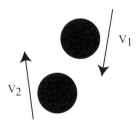

Figure 16.4. Two moving spheres colliding at a glancing angle.

total kinetic energy are conserved, but there is some transfer of both from one sphere to another.

Now, consider a reasonable volume of gas (cubic centimeters is a good way to picture it) in a sealed container, with all these billions of gas molecules crashing around in what we call Brownian motion, with the trajectories being the result of all of the elastic collisions. Since kinetic energy is conserved at each collision, the total kinetic energy of the gas molecules in the container is conserved (assuming of course that the collisions with the walls of the container are also elastic). Saying that the total kinetic energy of the gas molecules in the container is conserved is equivalent to saying that the average kinetic energy of the molecules is conserved, with this average just being the total divided by the number of molecules. There is a name for this average kinetic energy—we call it the *temperature* of the gas. This temperature may be measured in degrees Fahrenheit or Celsius, or in degrees Kelvin. Degrees Kelvin is a temperature scale that refers to 0 as the temperature at which the molecules no longer have any kinetic energy.[2] Zero degrees Celsius is equal to 273 degrees Kelvin, and the distance between degrees is the same in both systems.

Now, we have just mentioned an awful lot of "stuff." I have jumped from talking about the Brownian motion of gas molecules, through a discussion of elastic collisions of these gas molecules, to the concept of temperature as the average kinetic energy of the gas molecules. It turns out that these interacting molecules will *settle* on a distribution of energies among the molecules that remains constant (since we are dealing with billions of molecules in even a very small volume of gas, small random fluctuations are of no consequence).

I'll leave discussion of this distribution for a few paragraphs and concentrate on the picture of the gas molecules bouncing around in a (sealed) container. Let's put this container in outer space, so there are no gas molecules anywhere close outside the container. Figure 16.5 shows several of the gas molecules colliding with one of the side walls of the container. They hit the

---

[2] So long as we stay in the realm of classical physics, this is a valid concept. In actuality, the theory of quantum mechanics teaches us that even at zero degrees Kelvin, the so-called "absolute zero" of temperature, there is some activity.

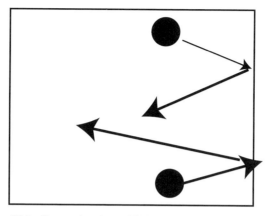

Figure 16.5. Gas molecules colliding with a wall of a container.

wall at random angles, but at a uniform distribution of angles. They ricochet off the wall and continue their motion (further collisions are not shown in this figure) about the container. What happens in one of these bounces is that the Y component of a molecule's momentum, parallel to the wall, doesn't change at all; the X component's momentum, however, reverses. According to Newton's law, therefore, these particles are pushing (exerting a force) on the wall.

Now Newton's law also tells us that when you exert a force on something, it starts to move. Is the box going to start to move to the right? The answer is a resounding no, because there is an equal number of molecules crashing into the left wall, and the forces balance out. If there are few enough particles and you look closely (and quickly) enough, you might detect a little vibration of the walls because of random variations in when the particles strike the walls. On the average, however, the forces cancel out and the box does not start moving.[3]

In a gas of a given density (gas molecules per unit volume), there will be an average number of molecules colliding with all surfaces per unit time. Rather than talk about the force on, say, the right-hand wall, it's more convenient to talk about the force per unit area of surface, which is called the pressure, on that surface. This makes sense because if we take a box with double the volume of our first box and put in double the number of gas molecules (at the same temperature), then we would expect to see the same number of collisions per unit area, and the pressure would be the same.

If I expand on the above argument a bit, I can easily generalize to the Ideal Gas Law. Let's take it in pieces. Start with a box of volume V containing N molecules of gas at temperature T. The resulting pressure on the walls of the

---

[3] Another way of expressing Newton's law about "a body at rest tends ..." is to say that a system may not change its own momentum. That's why a rocket ship has to pump "stuff" out the back—the total momentum of the system is constant, but the rocket ship goes forward.

box is P. I won't be able to actually derive the constant that relates all of these numbers in one set of units, but I will state the result of this calculation below. I'll call the scaling constant R.

Suppose we have a box of volume V at a temperature T. Due to the number of gas molecules in the box, the pressure exerted on the walls of the box due to the gas is P. Now, suppose we took the same number of gas molecules at the same temperature and put them in a bigger box, a box of volume $2V$. Since we have the same number of gas molecules as before but in twice the volume as before, the average density of particles is only 1/2 of what it was before. We would expect that in some given interval of time there would be only half the collisions with the walls as we had before, but that on the average the force of each of these collisions would be the same as it was before. The pressure on the walls of this bigger box would only be 1/2 what it was before. This is the same argument as getting a second box the same size as the original box but only putting in 1/2 the amount of gas.

Taking this one step further, if I doubled (or tripled, or ...) the size of the box while keeping the temperature and number of gas molecules constant, I would expect the pressure to drop to one-half (or one-third, or ...) of what it was before. Mathematically, this is written as

$$PV \propto n$$

where $\propto$ is notation for "is proportional to" and n is the number of particles.

The force exerted by the molecules on the wall is the product of the force per molecule and the rate of molecules hitting the wall. The pressure is then this force per unit area. The force per molecule is, by Newton's law, proportional to the rate of change of momentum. For a particle striking the right-hand wall and bouncing back into the box, as shown in Figure 16.5, the change of momentum per unit time is just the difference of the momentum just after and the momentum just before the collision with the wall. This in turn is directly proportional to $\mathbf{V}_x$ (with a similar statement about $\mathbf{V}_y$ on the upper and lower walls). Since there is a constant stream of molecules (on the average) hitting the wall, the rate of hits must be proportional to the speed of the molecules toward the wall, which again is just $\mathbf{V}_x$. Putting these pieces together, the force exerted by the molecules on the left and right walls is just $(\mathbf{V}_x) (\mathbf{V}_x)$ = $(\mathbf{V}_x)^2$. The force exerted by the molecules on the top and bottom walls, similarly, is $(\mathbf{V}_y)^2$. If we double this force while keeping everything else constant, we'll double the pressure.

Doubling the average $\mathbf{V}_x^2$ and $\mathbf{V}_y^2$ means doubling the average kinetic energy, which just means doubling the temperature.

Now we have all the pieces we need:

$$PV = nRT$$

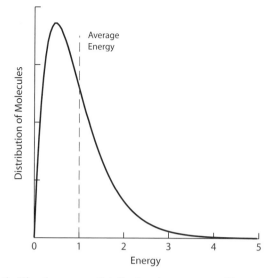

Figure 16.6. Kinetic energy distribution in a gas at uniform temperature.

where P is pressure; V is volume; n is the number gas molecules expressed in moles (One mole of a pure substance has a mass in grams equal to its molecular mass—that is, the sum of the atomic masses of the constituents of the molecule as found in the periodic table[4]); T is (absolute) temperature; and R, the constant that gets all the units right, is called the gas constant.[5]

This simple formula is called the Ideal Gas Law. It was derived using only the description of Brownian motion and the concepts of momentum, kinetic energy, and elastic collisions. Every chemistry student has seen this law, but the fact that it's just a conclusion reached from some statistical considerations is probably not recognized by many of them.

Now let me return for a moment to the picture of the Brownian motion of gas molecules in a box. No matter what the distribution of kinetic energies of molecules we start with, after the molecules have crashed around for a while (or "reached equilibrium"), the final distribution always looks the same. Figure 16.6 shows this distribution, for a gas with an average energy of 1.0 (arbitrary units). There is a noticeable tail to this function, with a small percentage of the molecules having 4 or 5 times the average kinetic energy any given time.

The air pressure at sea level is about 15 pounds per square inch. If you want to inflate a car tire, for example, then you must pump air into the tire so that the pressure inside the tire exceeds this (atmospheric) pressure by about 30 pounds per square inch. Since almost all common references to pressure only

---

[4] One mole contains the same number of particles as there are in 12 g of carbon-12 atoms by definition. This number is called *Avogadro's number*, which equals 6.022×10^{23} particles.
[5] R is 0.082057 L·atm·mol^{-1}·K^{-1}.

care about the relationship of this pressure to the atmospheric pressure, the term *gauge pressure* has come into common usage. Gauge pressure is the difference between the absolute pressure and the atmospheric pressure. Most common pressure gauges are calibrated in gauge pressure, even if they don't say so. For example, a car tire that's absolutely flat has zero *gauge pressure*. If you were to seal the valve and take this flat tire into outer space, it would look inflated and the amount of air in it would be just enough for the pressure on the inside walls of the tire to be 15 pounds/square inch while on the ground. If you were to pump air out of a stiff-walled canister so that the canister contained "a vacuum," the gauge pressure in the canister would be negative. This just means that the absolute pressure in the canister is less than atmospheric pressure—there is no such thing as a negative (absolute) pressure.

THERMODYNAMICS

If we were to mix together two gases at different temperatures, after a period of time the collisions would randomize things and the energy distribution would again look like Figure 16.6. The temperature, aka. the average kinetic energy, would be somewhere between the temperatures of the originally cooler gas and the originally hotter gas. The actual final temperature depends on both the original temperatures and the amounts (and types) of each gas. This result should not be surprising; it works with solids and liquids as well. When dealing with liquids, for example, if you mix hot water with cold water, you get lukewarm water. There's no way to end up with water that's either colder than the original cold water or hotter than the original hot water.

When the gases at different temperatures were separate, we had the opportunity to get some useful work (such as turning a turbine and generating electricity) out of them, as I shall demonstrate below. Once the gases are mixed together, however, even though no energy has been lost, the opportunity to get this useful work is lost. Furthermore, you would have to supply more energy to separate these gases again. This fundamental irreversibility of processes is related to a property called entropy which will be discussed a bit below.

The kinetic energy of a gas is, as the name implies, a form of energy. For a confined body of gas at equilibrium, the average kinetic energy is the temperature. The total kinetic energy is called the *heat energy* of the gas. The English language misleads us here, because the natural interpretation of saying that item A is hotter than item B is that item A contains more heat energy. Unfortunately, it doesn't mean that at all. "Hotter than" means "at a higher temperature than." As an example, consider a pot of warm water and an equal volume pot of boiling water. The boiling water is at a higher temperature and has more heat energy than does the bucket of warm water. You could casually dip your hand into the pot of warm water; but you wouldn't want to dip your hand into the pot of boiling water. Now consider a bathtub full of warm water, at the same temperature as the pot of warm water; but because of the significant difference in volumes, the bathtub of warm water contains more heat energy than does the pot of boiling water. Still, you could dip your hand into the bathtub of warm water without worrying about the consequences.

Historically, heat energy has been easy to come by. Burn something and you get heat energy, the energy coming from the energies holding together the chemical compositions being burnt.

Originally, heat was used to cook food and to keep people warm. Then it was discovered that heat energy could be used to do mechanical work. Figure 16.7 shows a very crude but workable heat-based machine. I burn something and use the heat to boil water; the boiling water produces steam which is, in the confined region shown, at a higher pressure than atmospheric pressure. The resulting "plume" of high-pressure steam escaping through the small orifice in my vessel spins a paddle wheel. The paddle wheel can push a boat through the water, turn a generator to make electricity, power a mill to grind flour, and so on.

This simple heat engine works because the heated steam is at a higher pressure than the atmosphere and pushes harder on one side of a paddle than the ambient air pressure is pushing back on the other side of the paddle. The hotter I can get the steam, the higher the pressure and the better the engine works. Because some work must always be expended to push the air

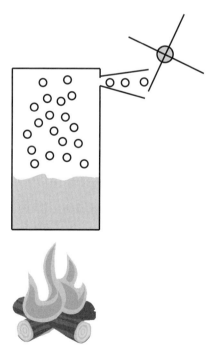

Figure 16.7. A simple heat engine.

away from the back side of the paddle, I will never get the amount of useful work out of the steam as I put energy into heating the steam. Since the back-pressure is a constant while the steam pressure is a function of the temperature that the steam is heated to, the higher the steam temperature, the smaller the fraction of energy wasted and hence the higher the efficiency of the engine.

These latter characteristics turn out to be very basic characteristics of heat engines. That is, you can never get full use of the energy you put into the system but you can increase the efficiency (the useful work you get for a given amount of energy put in) by increasing the temperature of the steam, or whatever it is you're heating as compared to the ambient temperature.

The study of thermodynamics has led to three fundamental laws that seem universal throughout the observed universe:

1. Conservation of energy—the total energy in the universe is constant. Of more practical value is that in a closed system—that is, a system in which energy cannot enter or leave—the total energy must be constant.

2. The entropy of any [closed] system that is not in equilibrium (not at a uniform temperature) will increase with time. Entropy is a measure of the disorder of the system. In the simple heat engine example above, if I generate the high-pressure steam and then just let it dissipate into the atmosphere, I haven't lost any of the heat energy that I put into the steam. Instead, I've cooled the steam a lot, heated the local air a little, and totally lost the ability to get any useful work out of the steam. So long as the hot steam and the cool ambient air are separated, we have an ordered system with a low entropy and an opportunity to get some useful work done. Once the hot steam and the cool ambient air are mixed together, we have a disordered system with a higher entropy and have lost the ability to get some useful work done (by the steam).

3. We can define an absolute zero of temperature as the temperature at which all thermal motion ceases. In general, cooling a system takes work and both the heat resulting from this and the heat removed from the system must be "dumped" somewhere else. Cooling a region lowers the entropy of the cooled region even though the entropy of the universe is increasing because of the effort. If a region could be cooled to absolute zero, its entropy would be zero. The third law says that it is impossible to reduce any region to absolute zero temperature in a finite number of operations.

Most students of this material who have not kept up with it lose familiarity with it over the years and could not easily quote these three laws to you. Virtually all of them, however, can quote the simple, everyday paraphrasing that is surprisingly accurate:

First Law of Thermodynamics: You can't win (i.e., you can't create "new" energy).

Second Law of Thermodynamics: You can't break even (i.e., you can't run any heat machine without wasting some of your energy).

Third Law of Thermodynamics: You can't get out of the game (i.e., you can never reach the zero entropy state of absolute zero temperature).

CHAPTER 17

INTRODUCTION TO STATISTICAL ANALYSIS

This book is mostly about probability, which I am defining to mean how to make best guesses about what is likely to happen in a situation with random variables controlling some, if not all, events as well as how close to my guess should I expect the results to be. Statistical analysis is the study of what you really know about a population distribution when you can't measure every member of the population and/or when your measurement capability has errors with their own probability distribution function(s). Statistical analysis has its mathematical roots in probability theory, so there can be no discussion of it without a prior discussion of probability. A discussion of probability theory, on the other hand, can stop short of going into statistical analysis, but since statistical analysis is probably the predominant customer for probability theory, I'd be remiss if I didn't at least introduce some of the major concepts.

One brief aside about the English language. I've used the common expression "look at the statistics" or similar phrases throughout the book when I meant "look at the PDF for ..." or "look at the histogram of. ..." The general term "statistics" is used to refer to just about anything that's got some randomness associated with it in some way. (I'm not being very precise here, but then again neither is the common usage.) Statistical inference, on the other hand, is defined a little more carefully, as I hope the material of this chapter demonstrates.

Probably Not: Future Prediction Using Probability and Statistical Inference,
by Lawrence N. Dworsky.
Copyright © 2008 John Wiley & Sons, Inc.

SAMPLING

Suppose we have a population of something that we want to learn about. The example used in Chapter 1 is the distribution of lifetimes of light bulbs. This example is mathematically equivalent to the average weight of bottles of juice, the average amount of sugar in bottles of soda, and so on. A related but not quite identical example is the incidence rate of broken cashew nuts in jars (of cashew nuts), or of nails without heads in boxes of nails. All of these examples are important to manufacturers (and purchasers) of these items.

Related to the above is the problem of measurement. No measurement instrument is perfect. In addition, measurements of weights and volumes tend to vary with ambient temperature and humidity, both of which are not easy to control precisely. A large manufacturer of some product will probably have multiple manufacturing lines running, each with their own measurement machines. Sometimes these machines are right next to each other in a factory, sometimes they're distributed in several factories around the world. There is an issue of correlation of the different measurement systems. Putting all of this together, if we were able to take one bottle of juice and measure its weight on every one of the manufacturer's scales in every factory, several times a day, in every season of the year, we'd find a distribution of results.

Lastly, in practice both of the above phenomena occur simultaneously. That is, every bottle of juice is slightly different from the others, and no two measurements of any single bottle of juice repeat exactly. How do we know what we have?

The first problem to address is the sampling problem. We don't want to run every light bulb until it burns out in order to see how long it lasts. We don't want to inspect every nail to see if it has a head. We would like to look at a small representative group of items and infer the information about the total population.

The last sentence above contained two seemingly innocuous words that actually each contain an entire universe of issues. What do we mean by *small*, and what do we mean by *representative*? We will be able to define *small* mathematically in terms of the size of the entire population, the size of confidence intervals, and so on. *Representative*, on the other hand, more often than not depends on judgment and is not easy to define. It is the factor that leads many surveys about peoples' opinions and preferences astray.

I'll start with an example that doesn't involve people. I'm a bottled juice manufacturer and I want to know the PDF of the weights of my bottles of juice without weighing every bottle. I could start out by taking a standard weight and using it to calibrate every scale in all of my factories, but the time involved in shipping this weight to my various factories around the world is large as compared to the time it takes for scales to drift out of calibration. Instead, I'll try to develop a set of standard weights and have a few of each located in every factory. I could have a revolving system, with some fraction of the weights from each factory being brought together and checked against one master, or "gold standard" weight.

How often do I need to calibrate my scales? This depends on the scales, the temperature and humidity swings in my factories, and so on.

Once I've calibrated my scales, I have to set up a sampling procedure. I could take, say, 10 bottles of juice at the start of each day and weigh them. The problem here is that my juice bottle filling machines heat up over the course of a day and the average weight at the end of the day isn't the same as the average weight at the beginning of the day. I'll modify my procedure and weigh three bottles of juice from each filling machine three times a day. This is getting better, but my sample size is growing. As you can see, this is not a trivial problem.

The problems involved with surveying people are much more involved. Suppose I want to know the average number of hours that a person in New York City sleeps each night and what the standard deviation of this number is. I could start with a mailing list or a telephone list that I bought from, say, some political party. It's probably reasonable to assume that Democrats, Republicans, libertarians, anarchists, and Communists all have statistically similar sleep habits. Or do they? I don't really know, so I'm already on thin ice buying this mailing list.

I know that I won't get responses from everybody. If I telephone people, then I could pick up a bias; people who aren't sleeping well are probably crankier than people who are sleeping well and are more likely to hang up on me. Also, should I call midday or early evening? People who work night shifts probably have different sleep patterns than people who work day shifts, and I'll get disproportionate representation of these groups depending on when I call.

Lastly, how well do most people really know how many hours they sleep every night? Our memories tend to hold on to out-of-the-ordinary events more than ordinary events. If I had a cold last week and didn't sleep well, I'll probably remember this more than the fact that I slept perfectly well every night for the previous month. Also, studies have shown that many people who claim that they "didn't sleep a wink last night" actually get several hours of sleep in short bursts.

Finally, although it's hard to imagine that people would want to lie about how well they're sleeping, in many situations we simply have to consider the fact that people are embarrassed, or hiding something, and do not report the truth. Sexual activity and preference studies, for example, are fraught with this issue. Also, mail or email survey responses are usually overrepresented by people with an ax to grind about an issue and underrepresented by people who don't really care about the issue.

The bottom line here is that accurate and adequate sampling is a difficult subject. If you're about to embark on a statistical study of any sort, don't underestimate the difficulties involved in getting it right. Read some books on the subject and/or consult professionals in the field.

One comment about sampling before moving on. Many people and companies get around the problems of good sampling by simply not bothering. A good example of this is often seen in the health food/diet supplement industry. Valid studies of drugs, including health foods and "disease preventative" vitamins and diet supplements, take years and include many thousands of people. There are so many variables and issues that there really isn't a simple shortcut that's valid. Instead, you'll see bottles on the shelves with testimonials. Good luck to you if you believe the claims on these bottles. For some reason, many people will believe anything on a bottle that claims to be Chinese or herbal, but suspect the 10-year, 10,000-people FDA study that shows that the stuff is, at best, worthless. In the case of magnetic or copper bracelets or cures for the common cold, this is relatively harmless. In the case of cancer cures, you could be signing your own death certificate.

Returning to the mathematics of sampling, the most bias-free sampling technique is random sampling. In the case of the bottled juice manufacturer, use a computer program that randomly picks a time and a juice bottle filling machine and requests a measurement.[1] Some insight must be added to prescribe the average number of samples a day, but this can usually be accomplished with a little experience and "staring at" the data.

An assumption that we must make before proceeding is that the statistics of the population of whatever it is we're studying is not changing over the course of our study. Using the juice bottle example from above, suppose we're weighing 25 bottles of juice at randomly chosen times each day, but a worker (or a machine glitch) resets the controls on the filling machine every morning. All we're accomplishing in this situation is building a history of juice bottle weighings. The information from Monday is of no value in telling us what we'll be producing on Tuesday.

SAMPLE DISTRIBUTIONS AND STANDARD DEVIATIONS

Consider a factory that produces millions of pieces of something that can easily be classified as good or defective. The example of nails, where missing heads comprise broken parts, fits this category. There is a probability p of the part being defective and $(1 - p)$ of the part being satisfactory to ship to a customer. The manufacturer plans to allow the shipment of a few defective parts in a box along with thousands of good parts. The customers will do the final sorting: They'll pick a nail out of the box and if it's bad, and they'll toss it away. So long as the fraction of defective parts doesn't get too large, everybody is happy with this system because the quality control/inspection costs are very low. In principle, the manufacturer could inspect every part and reject

---

[1] Even this can go wrong in the real world. For example, a well-meaning worker might try to make his day simpler by grabbing a dozen bottles of juice in the morning, putting them near the scale, and then weighing one of them at each time prescribed by the computer program.

the bad ones, but this would be very expensive. On the other hand, if p is too large, the customers will get annoyed at the number of bad parts they're encountering and will switch to another manufacturer the next time they buy their nails. It is important for the manufacturer to know p accurately without having to inspect every part.

The manufacturer decides that every day he will randomly select n parts from the output of his factory and inspect them. If the parts are fungible and are sold in boxes, such as nails, then a box of n nails is a likely sample. On Monday he finds x bad parts and calculates the probability of a part being bad as

$$\bar{p} = \frac{x}{n}$$

What we want to know is p, the probability of a bad part (or, equivalently, the fraction of bad parts) that the factory is producing. We have not measured p. We have measured the fraction of bad parts in a random sample of n parts out of the population distribution. I have called this latter number $\bar{p}$ (usually read "p bar") to distinguish these two numbers. p is the fraction of bad parts in the population, and $\bar{p}$ is the fraction of bad parts in the sample of the population (usually simply referred to as the sample).

The manufacturer would like these two numbers to be the same, or at least very close, by some measure. We can get a feeling for the problem here by looking at using the binomial probability formula. Suppose that $p = 0.15$, so that $(1 - p) = 0.85$. This defective part fraction is undoubtedly much too large for a real commodity product such as a box of nails, but is good for an example.

We would expect to find about 15% bad parts in a sample. The probability of finding exactly 15 bad parts in a sample of 100 parts is

$$C\binom{n}{k}p^k(1-p)^{n-k} = \frac{100!}{85!15!}(.15)^{15}(.85)^{85} = 0.111$$

There is about an 11% probability that $\bar{p}$ will correctly tell us what p is. Figure 17.1 shows the probability of getting various fractions of bad parts in a random sample of 100 parts. As can be seen, the mean of the curve is at 0.15. The curve looks normal. This shouldn't be a surprise because several chapters ago I showed how a binomial probability will approach a normal curve when the n gets big and p or q isn't very large. The standard deviation is 0.0395. The 95% confidence interval is therefore 0.079 on either side of the mean. This means that we can be 95% confident that we know the fraction of bad parts being produced to within 0.079 on either side of the nominal value (0.15).

Also shown in Figure 17.1 is the results of doing the same calculation for $n = 200$. Again the mean is 0.15, but now the standard deviation is 0.0281.

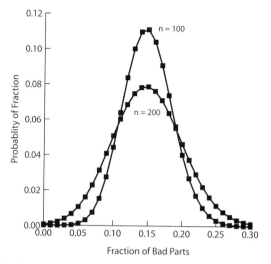

Figure 17.1. PDF of fraction of bad parts measurement for $p = 0.15$, $n = 100$ and 200.

Since the standard deviation has gotten smaller, we have a better, smaller confidence interval and hence a better estimate of the fraction of bad parts the factory is producing. If I take the ratio of the two standard deviations, I find that the larger number is 1.41 times the smaller number. This is, not coincidentally, a ratio of $\sqrt{2}$. The standard deviation of mean of the sample distribution falls with the square root of the size of the sample.

The standard deviation of $\bar{p}$ is given by

$$\sigma = \sqrt{\frac{p(1-p)}{n}}$$

One interesting point about the above equation: p is the probability of a defective part, so $q = 1 - p$ is the probability of a good part. The equation has the product of these two probabilities; that is, its value does not change if we put in the value of q rather than the value of p: $p(1 - p) = q(1 - q)$. The standard deviation of the fraction of good parts in a sample of size n is the same as the standard deviation of the fraction of defective parts in a sample of the same size.

ESTIMATING POPULATION AVERAGE FROM A SAMPLE

Next let's look at a population with a normal distribution of some property. The weights of bottles of juice is a good example. A factory produces bottles of juice with an average weight (mean of the distribution) w and a standard deviation σ.

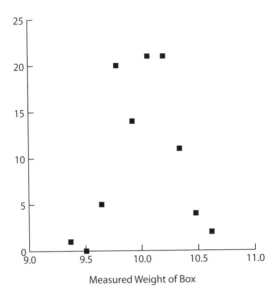

Figure 17.2. PDF of 250 ideal measurements of a population with mean = 10.0, σ = 0.25.

Again, for the time being, I'll assume that our measurement system (scales) is ideal.

We randomly select *n* bottles of juice and weigh them. Naturally, we get *n* different results. Figure 17.2 shows a histogram of a possible distribution of 100 measurements of a population with $W = 10.00$ and $\sigma = 0.25$. The average of these numbers is 9.986 and the standard deviation is 0.247. This sample seems to emulate the true population in average and standard deviation, but the histogram of the data certainly does not look like a well-behaved normal distribution.

In order to see how well averaging the results of a 100-measurement sample predicts the mean of the population, I'll repeat the procedure 100,000 times and graph the results (Figure 17.3). The average I obtained is 9.9999 and the standard deviation is 0.025. Remember that this average is the mean of a very large number of 100-measurement averages. This mean is a random variable itself, with an expected value of *W* and a standard deviation of

$$\frac{\sigma}{\sqrt{n}}$$

The average of *n* samples is the best estimate of the mean of the population, and the standard deviation of this average decreases as *n* gets larger. In other words, to get a very good idea of *W*, take as many randomly sampled measurements as you can and average them. Unfortunately, since the standard

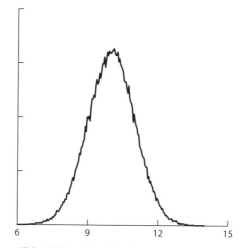

Figure 17.3. PDF of the distribution of $n = 100$ samples.

deviation of your result is only falling with the $\sqrt{n}$ rather than with n itself, improvement gets more and more difficult as you proceed. Going from $n = 25$ to $n = 100$, for example, will cut the standard deviation in half. To cut it in half again, however, you would have to go from $n = 100$ to $n = 400$.

If the measurement system has some variation, then we must somehow take this variation into account. Suppose we took one bottle of juice and measured it many times by randomly selecting the different scales in our factory(ies), different times of day, and so on. The average of all of our results is our best estimate of W, the weight of this bottle of juice. If, instead of a bottle of juice, we used a standard weight, then we would calibrate our system so that this average of all of our results is indeed the weight of the standard.

For a measurement system standard deviation σ_m and a population standard deviation σ, the total variation we have to deal with is

$$\sigma_{\text{tot}} = \sqrt{\sigma^2 + \sigma_m^2}$$

assuming that these two standard deviations are independent (of each other). Again, the standard deviation of the average of the measurements of W will get better with more samples:

$$\frac{\sigma_{\text{tot}}}{\sqrt{n}}$$

The formulas above assume that the population you're studying is normal, or nearly normal. This is usually a pretty good assumption, but be a little suspicious.

Also, these formulas assume that you know σ, the standard deviation of the population, and, if necessary, σ_m, the standard deviation of the measurement repeatability. What if you don't know these numbers? The easiest way to approximate them is to take the standard deviation of the n samples. Just in case some confusion is building here, remember that the standard deviation of the n samples in a measurement set must approach the standard deviation of the population as n gets large and eventually approaches the size of the population itself. The standard deviation that gets smaller with increasing n is not something you usually measure—it's the standard deviation of the averages of a large number of n-sized measurement sets.

THE STUDENT T DISTRIBUTION

Intuitively, it's easy to picture a set of a few hundred measurements having a standard deviation that approximates the standard deviation of the entire population. But what about a dozen measurements, or five, or even just two? With just one measurement, we can't calculate a standard deviation, so two measurements is the smallest set of measurements that we can consider when talking about approximating a standard deviation. Remember that the calculation of standard deviation has the term $n - 1$ in the denominator of the formula, and this makes no sense when $n = 1$. The $n = 2$ case is the smallest size measurement set possible and is referred to as a "one degree of freedom" case.

The best way to look at what happens with small values of n (small samples) is to *normalize* the distribution curves—that is, to first shift the curves on the horizontal axis until they all have the same mean, then to divide by $\sigma/\sqrt{n}$. In texts and handbooks it is common to define a "normalized" random normal variable z:

$$z = \frac{\bar{x} - \mu}{\sigma/\sqrt{n}}$$

This transformation makes it possible to calculate properties of all normal distributions by just studying one normal distribution.

Keeping the letter σ as the population standard deviation, let's differentiate the standard deviation calculated from the n-sized sample by calling it s. Then we'll define a new normalized variable as t:

$$t = \frac{\bar{x} - \mu}{s/\sqrt{n}}$$

Figure 17.4 is a repeat of Figure 17.3, but for the variable t using samples $n = 2$, $n = 3$, $n = 10$, and for the variable z. The first three of these correspond to 1, 2, and 9 degrees of freedom, respectively.

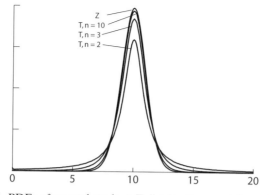

Figure 17.4. PDFs of several student *T* distributions and the *z* distribution.

The *T* curves are usually referred to as "Student *T*" distributions, after the mathematician who developed them using the pen name Student. As *n* increases, the Student *T* curves look more and more normal until at about *n* = 30 they're virtually indistinguishable (from the normal distribution). For low values of *n* (small samples), however, the curves are flatter and wider than normal curves. For a normal distribution, the 95% confidence interval is 2 standard deviations on either side of the mean. For an *n* = 2, or 1 degree of freedom, Student *T* distribution, the 95% confidence interval is about 12.7 standard deviations on either side of the mean.

If distributions had personalities, we would say that the Student *T* distribution is a very suspicious PDF. If you don't have too much information (i.e., a small sample), then your calculation of *s* is only a poor approximation of σ, and your confidence interval is very wide. As *n* increases, the confidence interval decreases until it reaches the value predicted by the normal curve.

One last word before moving on. What if the original population (e.g., the juice bottle weights) is not normally distributed? It turns out that it doesn't really matter: The Central Limit Theorem guarantees that the distribution created by taking the average of *n* experiments many times and calling this average our random variable will be normally distributed. The conclusions reached above are therefore all valid.

Converting our calculations to the *z* or *t* parameters offers us the convenience of seeing everything in a common set of parameters. That is, we are transforming all normal distribution problems to that of a normal distribution with a mean of 0 and a σ of 1. We know that the range between *z* = −1.96 and *z* = +1.96 is our 95% confidence interval, and so on.[2] Figure 17.5 shows the 90%, 95%, and 99% confidence interval points. Up until about 1985, it was

---

[2] It is common practice to refer to the normal distribution's 95% confidence interval as ±2σ even though ⊥1.96σ is more accurate.

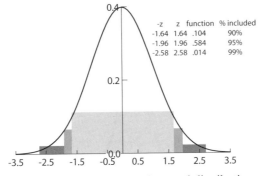

Figure 17.5. Z-transformed normal distribution.

necessary to carry around tables of all of these values, but personal computers and modern spreadsheet and statistical analysis software have relieved us of this need.

POLLING STATISTICS

If you are reading this book any time when there's a local or national election on the horizon, you are no doubt satiated with pre-election polling results in the newspapers and the TV and radio news reports. In a city of 1 million voters, for example, a pollster sampled 1000 people (hopefully randomly selected) and found that 52% favored candidate A. This doesn't seem like a large margin. Exactly how comfortable should candidate A feel the night before the election? This falls back on seeing how confident we are that the sample probability of 52% of the voters voting for candidate A represents the entire population of 1,000,000 voters.

We estimate the standard deviation of the sample data as

$$s = \sqrt{\frac{\bar{p}(1-\bar{p})}{n}} = \sqrt{\frac{0.52(1-0.52)}{1000}} = .0158$$

For a 95% confidence factor, we need 1.96 standard deviations, or $(1.96)(.0158) = 0.031$. Our confidence interval on the voting is therefore 0.52 ± 0.031. We are 95% confident that candidate A will get between about 49% and 55% of the vote. Remember that if we had taken more samples, n would be larger and s would be smaller than the values shown above. The confidence interval about the 52% result of the survey would have been smaller, and we could be more comfortable with the prediction.

The above example, while mathematically correct, can easily be misinterpreted. I showed that the 95% confidence window about the 52% sample result is ±0.031. Candidate A, however, doesn't really care about the + side of things; his only concern is that the actual vote count might give him less than 50%. We should therefore ask just what confidence factor we have of our candidate winning. We need a confidence interval on the low side of 0.52 − 0.02 to keep him in winning territory. Therefore we need to calculate 0.02 = $(z)(.0158)$, or $z = 0.02/0.158 = 1.27$. Going to a spreadsheet or a table, I find that this is just about an 80% 2-sided confidence factor; that is, there is a 10% probability of getting less than 50% of the vote and a 10% probability of getting more than $0.50 + 0.02 = 52\%$ of the vote. Since the relevant information for our candidate is only the low side, he has a 90% confidence factor of winning. This isn't a guaranteed win, but it's not too bad for the night before the election.

DID A SAMPLE COME FROM A GIVEN POPULATION?

Another use for confidence factor calculations is to determine how likely it is that a sample data set really came from a random sampling of some population. As an example, consider left-handedness. The national incidence of left-handedness is about 12% of the total population. In a randomly selected group of 33 people we should expect to find about 4 left-handed people. But what if we stand on a street corner, query the first 33 people we meet, and find 7 left-handed people? Are we within the expected bounds of statistical variation or should we be suspicious about this sample?

I'll start by calculating the sample standard deviation:

$$s = \sqrt{\frac{p(1-p)}{n}} = \sqrt{\frac{0.12(1-0.12)}{33}} = 0.057$$

We would expect $\bar{p}$ to be within $1.96s$ of p (on either side) for a 95% confidence factor. Calculating the number, $p + 1.96s = 0.12 + 0.112 = 0.232$.

From the sample data, $\bar{p} = 7/33 = 0.212$. Therefore, our sample data are within the confidence interval. To within this confidence interval, we can therefore conclude that finding 7 left-handed people in a group of 33 isn't suspicious.

What if our sample is a small group? If we only queried 11 people, we would expect to find 1 left-handed person. At what point should we get suspicious? This problem is worked out the same as above, except that we must use the Student T distribution. Going to the tables or to any appropriate software, we find that for a 10 degree-of-freedom ($n − 1 = 11 − 1 = 10$) Student T distribution, instead of the 1.96σ number used for the normal distribution, we should use 2.23σ. Without repeating the arithmetic, I'll just state the conclusion: Anywhere from 0 to 3 left-handed people are OK to a 95% confidence factor limit. For a group of only 2 people, we must use 12.7σ. We might intuitively feel that we shouldn't see any left-handed people in a group of only 2, but seeing 1 left-handed person in our group of 2 "just misses" the 95% confidence interval; and while it's therefore pretty unlikely, it shouldn't be a major cause for alarm.

There's much, much more to the field of statistical analysis than has been presented. However, as I said at the beginning of this chapter, this is primarily a book on probability, so I'll stop here. Hopefully, I've conveyed the flavor.

CHAPTER 18

CHAOS AND QUANTA

I'm going to close with short discussions of two enormous subjects.

Chaos theory has been popularized by the concept that "If a butterfly flaps its wings in Brazil, it can set off a tornado in Texas." I'm going to take a quick look at some of the ideas here because chaotic data can sometimes look like random data. In many cases it can be successfully used instead of random data for Monte Carlo simulations and the like. Chaotic data are, however, totally deterministic. I think that showing how to contrast chaotic data and random data can be interesting.

Quantum mechanics is one of the masterpieces of twentieth-century physics. Assuming that I was capable of providing it, a full description of quantum mechanics would take many thick texts to contain it and would require an extensive mathematics and physics background to fully understand it. As we understand it today, quantum mechanics best explains how the universe works. At the heart of quantum mechanics is the idea of truly random events—not "effectively random" like the flip of a coin, nor pseudorandom like the numbers pouring out of a computer algorithm, but truly, truly random.

CHAOS

Figure 18.1 shows what the output of a (pseudo)random number generator, putting out numbers between 0 and 1 in a row, looks like. Compare this to Figure 18.2, which contains 100 values of the function

Probably Not: Future Prediction Using Probability and Statistical Inference,
by Lawrence N. Dworsky.
Copyright © 2008 John Wiley & Sons, Inc.

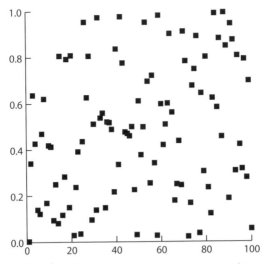

Figure 18.1. One hundred points from a random number generator.

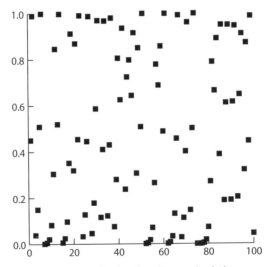

Figure 18.2. One hundred points from a logistic map, $a = 4$.

$$x_{\text{new}} = ax_{\text{old}}(1 - x_{\text{old}})$$

run recursively.[1] For this figure, a = 4 and the formula was run 500 times before the 100 values to be graphed were recorded. This function is known as a "Logistic Map" generator.

[1] "Recursively" means feed the output of a function back into the input and get a new output, and so on.

Both of these figures look pretty random. Both of the lists have (approximately) the correct mean and sigma values for a set of random numbers between 0 and 1. Is there really a difference or is the above equation a very easy way to generate pseudorandom numbers? The generated list of numbers doesn't seem to show any simple repetition patterns, or anything else obvious.

Figure 18.3 uses the same list of numbers used in Figure 18.1 but is shown in what is called phase space. This graph is generated by taking each number on the list (except the last) and using it as the X, or horizontal axis, value and then taking the next number on the list and using it as the Y, or vertical axis, value. Table 18.1 shows how this works for the first 11 numbers on the list.

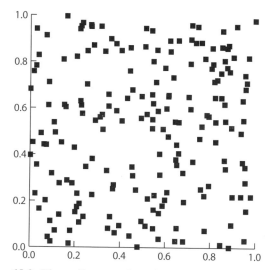

Figure 18.3. Phase diagram of random number generator data.

TABLE 18.1. How a Phase Plot Is Generated

List of Numbers	Plotted Points	
	X	Y
0.450432	0.450432	0.990170
0.990170	0.990170	0.038935
0.038935	0.038935	0.149676
0.149676	0.149676	0.509092
0.509092	0.509092	0.999667
0.999667	0.999667	0.001332
0.001332	0.001332	0.005321
0.005321	0.005321	0.021172
0.021172	0.021172	0.082893
0.082893	0.082893	0.304087
0.304087		

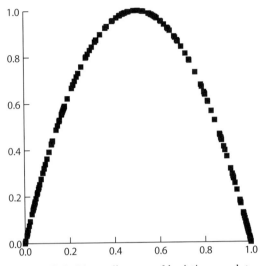

Figure 18.4. Phase diagram of logistic map data.

This graph doesn't really look different from the first two graphs of this chapter—everything still looks pretty random.

Figure 18.4 is the same as Figure 18.3, but the list of numbers from the logistic map used in Figure 18.2 is used. This figure is clearly different from the others: It certainly does not look like a plot of random combinations of numbers.

As I mentioned above, chaos theory is a sophisticated, involved branch of mathematics that certainly can't be summarized in a few paragraphs. Being that I've introduced the topic, however, I'll try and present a little perspective of what's going on and why people care.

Let me go back to the *logistic map* equation above. Figures 18.5a–c show what this (recursive) equation produces for 50 recursions using 3 different values of the parameter a. In each case that I started, or *seeded*, the equation with the value $x = 0.5$. For $a = 2.5$ (Figure 18.5a), x quickly settles to a fixed value of 0.6. This is not exciting. For $a = 3.5$, x never settles to a fixed value, but rather takes on what is called *periodic* behavior. That is, it repeatedly cycles between (in this case, 4) fixed values.

At $a = 3.75$, all hell breaks loose. This is the type of behavior shown earlier that looks random until you draw a phase plot.

The behavior of a function such as the logistic map can be summarized in what is called a *bifurcation diagram*, shown in Figure 18.6. I generated this diagram by taking successive values of the parameter a; for each value of a, I ran the recursive function (the logistic map) 100,000 times just to make sure that I was giving things a chance to settle down if they were going to settle down, and then I ran the recursion function 50 times more and plotted the resulting values of x.

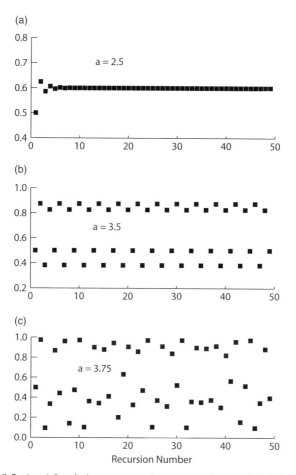

Figure 18.5. (a–c) Logistic map equation output for a = 2.5, 3.5, and 3.75.

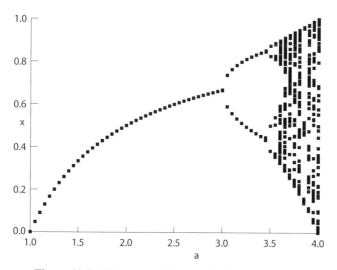

Figure 18.6. Bifurcation diagram of the logistic map.

For a less than about 3, the 50 values of x (at each value of a) are identical, and the points lie right on top of each other. For a between about 3 and 3.4, there is a simple periodicity: The values of x bounce back and forth between a high value and a low value. For $a = 3.5$ we see the slightly more complex periodicity described above: The values of x bounce back and forth between 4 different values. Then, finally, for a greater than about 3.7 we see the truly chaotic behavior of this function.

Now I can return to the question of why we care.

Figure 18.6 illustrates the "Butterfly Wing" effect. A very slight change in a causes drastic changes in both the number of values of x and the values themselves. I didn't show the results for a between 0 and 1 because these results were very unexciting—they were identically always 0. Now, what if a much more complex equation or system of equations had similar behavior. Let's say that a is related to the average temperature and that x is related to the peak winds of hurricanes around the world in a given year.[2] If a is small enough, there are no hurricanes that year. If a is a bit bigger, there will be 1 or 2 storms. If there are 2 storms, 1 of them will be moderately strong. If a is a bit bigger yet, there will be an unpredictable number of storms with wildly varying intensities. Furthermore, in this latter chaotic region, the subtlest change in a can cause drastic changes in both the number and intensities of the storms.

Another implication of chaotic behavior is referred to as *fractal behavior*. Look at Figure 18.7a. The x values represent the input to a function that describes some sort of physical behavior. Let's say it's the speed that you're throwing a ball, straight upward. Then the y value could represent the height the ball reaches. Looking at the figure, if I throw it with a speed of 4 (nonsense units), then it reaches a height of about 14. If I throw it with a speed of 5, then it reaches a height of about 17 units. Because the graph describing this function is a straight line, the relationship is called linear. If I want to know what will happen if $x = 6$, then I can extrapolate by looking at the line (and get $y = 20$).

Now look at Figure 18.7b. Whatever physical process I'm describing, it's certainly not a linear response. I cannot, for example, look at the response (y value) to $x = 4$ and the response to $x = 6$ and gain any insight as to the response at $x = 7$.

However, I can say that if I keep narrowing the gap between the two values of x, I will ultimately get to a region where I can estimate the value of the response at a value of x slightly higher than my larger value of x just as I did in the linear case. In Figure 18.7b, look at the points shown, $x = 4.3$ and $x = 4.35$; I can get a pretty good guess of the value of y for $x = 4.4$.

For many years it was assumed that the world around us worked in this manner. If x is time, then we are trying to predict the response of some physi-

---

[2] This is certainly not real meteorology. It's just a hypothetical example to demonstrate the possibilities and implications of chaotic behavior.

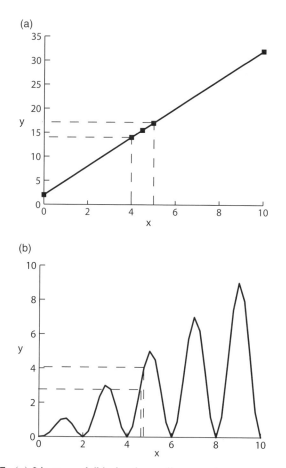

Figure 18.7. (a) Linear and (b) simple nonlinear system response examples.

cal system in the future by gathering a lot of very detailed, closely spaced data of the system up until the present. If we could get information about the temperature and barometric pressure and wind velocity at very closely spaced positions around the earth every (say) minute, then we should be able to do some really good weather forecasting.

Now look at Figure 18.8. Figure 18.8a is a function plotted between $x = 0$ and $x = 100$. This looks like a mountain viewed from a distance. I want to estimate the value of the function at $x = 50$ based on measurements at either end of the graph, in this case $x = 0$ and $x = 100$. Clearly, I can't do this, so I'll "zoom in," as if I had a set of binoculars with a magnification of 2, and obtain Figure 18.8b. I don't seem to be any closer to an estimate of the value at $x = 50$, so I'll try to look a little closer. Figure 18.8c shows a "zoom in" of a factor a 4 from the original (Figure 18.8a). Now I'm looking between $x = 37.5$ and $x = 62.5$, and my knowledge of the value of the function at these points doesn't

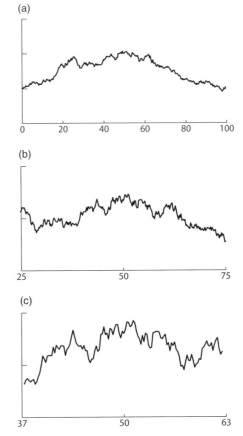

Figure 18.8. View of a fractal landscape at 1×, 2×, and 4× zoom levels.

help me at all to estimate the value at $x = 50$. The more I zoom in, the less I seem to be able to estimate.

This type of "fractal behavior" actually happens when you look at a mountain. When you look at a car, assuming it's not a trick car manufactured just to fool you, you know about how big the car should be, and this information helps your brain to process what you see and determine how far away the car is. When you look at a big mountain very far away (e.g., Figure 18.8a) or a small mountain up close (e.g., Figure 18.8c), you see about the same thing— and your brain cannot help you to interpret what you're seeing in terms of how far away it really is. Also, suppose I want to know about how many foot-steps it will take to walk along the top of this mountain. Figure 18.8a will give me an approximate answer; but by looking at Figures 18.8a and 18.8c, I see that the walk will actually be a lot longer than I had first thought.

The relevance of all of the above to the subject of probability theory is that we have measurements and observations that look as if there is an element of randomness involved, but that's not the case at all. Both the information presented in a bifurcation diagram and looking at fractal patterns are completely deterministic—there is no element of randomness involved at all. This is completely contrary to many of the discussions in the book where there is a tendency to see deterministic patterns when in fact we're dealing with randomness. It explains, among other things, why there is a limit to how good long-term weather forecasting can ever be.

PROBABILITY IN QUANTUM MECHANICS

Some atoms are called *radioactive*. This means that they emit particles (electrons or helium nuclei) or electromagnetic radiation (gamma rays) and the atoms change to another type of atom. This process is known as *radioactive decay*. As an example, assume we have a material whose atoms each has a probability of 0.01 (1%) of radioactive emission per day. Let's start with 10^{23} of atoms that emit electrons. This might seem like a very large number (compared to most of the things we actually count every day, it *is* a very large number), but it is approximately the number of atoms in a small lump of material.

Since we're dealing with such a large number of atoms, we can treat the expected value of atoms emitting as just what's going to happen. On the first day, this material would emit $(.01)(1 \times 10^{23}) = 1 \times 10^{21} = 0.01 \times 10^{23}$ electrons, and this same number of particles would no longer be radioactive. There would be $1 \times 10^{23} - 0.01 \times 10^{23} = 0.99 \times 10^{23}$ atoms of radioactive material left.

On day 2, therefore, the material would emit $(.01)(0.99 \times 10^{23}) = 0.0099 \times 10^{23}$ electrons and there would be $0.99 \times 10^{23} - 0.0099 \times 10^{23} = 0.9801 \times 10^{23}$ atoms of radioactive material left. Figure 18.9 shows the fraction of radioactive material left as a function of time for about 2 years. For those of you interested in the math, this figure is identically the curve[3]

$$y = e^{-.01x}$$

If you're not interested in the math, don't worry about it. I should note, however, that although we are only dealing with a small probability each day for a given atom, since we're dealing with so many atoms at once, the standard deviation in the number of atoms decaying in, say, the first day is

$$\sigma = \sqrt{np(1-p)} \approx \sqrt{np} = \sqrt{\mu} = \sqrt{1 \times 10^{23}} \approx 3 \times 10^{11}$$

---

[3] This function is known as *exponential decay*. It describes any situation where the rate that you're losing something (it's decaying) is directly proportional to the amount you still have. *Exponential growth* would be represented by the same equation without the minus sign in the exponent; in this case the rate that something is growing is directly proportional to the amount you already have.

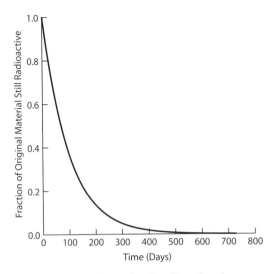

Figure 18.9. Example of radioactive decay.

This is a very large number, but as a fraction of the mean (np), it's just $3 \times 10^{11}/1 \times 10^{23} \sim 3 \times 10^{-12}$, which is a very small number. This is why I said that, even though the "performance" of an individual atom is very erratic, the observable decay rate of a small lump of material is a very smooth curve.

After approximately 69 days, half of the starting material is still radioactive. This material is therefore said to have a "69-day half-life."[4] Actual material half-lives vary wildly. For example, one form of astatine has a half-life of ~0.03 seconds and one form of beryllium has a half-life of over a million years!

Having characterized radioactive decay, physicists then started asking, What prompts an individual atom to decide to decay? The amazing answer, couched in very unscientific terms, is that "it decays when it wants to." That is, this very basic characteristic of some types of atoms is a truly random process. It is governed by the laws of probability as a very fundamental fact of nature. We have here a phenomenon that doesn't just *look* random, such as the output of a pseudorandom number generator, but really *is* random, with a PDF for radioactive decay that is as much a part of the fundamental nature of this atom as the number of electrons orbiting the nucleus.[5]

The concept that random events are at the very core of the mechanics of the universe was not accepted readily when it evolved in the early twentieth century. Albert Einstein, one of the key contributors to the early quantum theory, purportedly never accepted it and left us with the famous quote that "God does not play dice."

---

[4] The actual relationship is half-life = Ln(2)/p ~ .693/p where p = rate of decay per unit time.
[5] It's not hard to picture how a radioactivity counter and a small piece of radioactive material could be combined to form a true random number generator.

One of the paths investigated to look for determinism rather than random-ness was the idea of *hidden variables*. This is directly analogous to the opera-tion of a pseudorandom number generator, but looking backwards: A good pseudorandom number generator will produce a sequence of numbers that pass all your tests for randomness until someone shows you the algorithm that produced the numbers. Then you realize that you have been dealing with a totally deterministic sequence of numbers that just looked random. Physicists pursuing this path hoped to extend their understanding of atomic physics to a deeper level that would reveal some deterministic rules that just happened to produce results that looked random. They never succeeded.

The name quantum mechanics came about because of the observation that electrons in atoms cannot take on any level of energy. The energy can only be certain discrete, or quantized, values. This, along with the observation of truly random events, is totally different from classical, or Newtonian, mechanics.

Another observed phenomenon that has no counterpart in classical mechan-ics is tunneling. In order to describe tunneling, I have to first introduce the idea of a *potential well*.

Suppose you're lying on the floor in an open-air room (it has no ceiling) with walls 15 feet high. Lying on the floor isn't an essential part of the physics, but it's convenient in that anything in your hand is essentially at ground level. You have a rubber ball in your hand that you throw, your intent being to have the ball fly out over the top of the wall and leave the room. When you throw the ball, you are imparting kinetic energy to the ball. As it climbs, the ball is working against the downward force of gravity and the kinetic energy is being transferred to potential energy. Since the kinetic energy is proportional to the (square of the) speed, the ball is slowing down. When the kinetic energy reaches 0, the ball stops climbing and, under the influence of gravity, starts falling.

The height that the ball reaches before falling is related to how "hard" you throw the ball—that is, its starting speed. If the starting speed isn't fast enough, the kinetic energy isn't large enough and the ball will never clear the top of the wall. This is illustrated in Figure 18.10 (just the ball, I've omitted you lying on the floor) for two different initial velocities.

The point here is that in this classical, or Newtonian, physics picture of the world, a ball with insufficient initial kinetic energy well never leave the box. The total energy of the ball, its kinetic energy plus its potential energy, is constant. We can therefore say simply that the ball must have a certain minimum total energy in order to escape the box. Note also that the width (thickness) of the wall is totally irrelevant in this discussion.

The energy required for a ball to escape the box is directly related to the height of the walls. If we wanted to, we could describe the height of the walls in terms of this required energy rather than their height. This seems awkward when discussing throwing balls over walls, but it's perfectly valid.

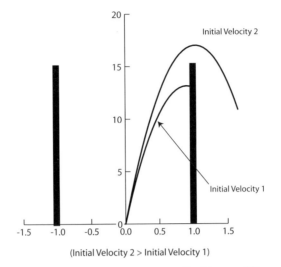

(Initial Velocity 2 > Initial Velocity 1)

Figure 18.10. Trajectories of balls thrown at a wall with two different energy levels.

In a metal, electrons of sufficient energy can move around freely. This is, in a sense, analogous to a bouncing ball rattling around in a box. There is an energy barrier that keeps the electrons from leaving the metal—again analogous to the ball rattling around in a box. Furthermore, if the metal is heated, some electrons gain enough energy to overcome the barrier and escape the metal. This is called thermionic emission and is the basis of electronic vacuum (radio) tube operation.

Quantum mechanically, the position of the electron can be described as a wave. The mathematical description of this wave is the solution to the equation developed by Erwin Schrödinger and which carries his name. The physical interpretation of this wave is that the probability of the location of the electron is monotonically related to the amplitude of this wave.[6]

Figure 18.11 shows the wave function of an electron in a box confined by potential energy walls of finite height and thickness. This is a one-dimensional representation of what is actually a three-dimensional situation. The wave function is largest in the box, indicating that it's most likely that the electron will stay confined. The wave function, however, is continuous across finite-height energy barriers. It falls off quickly with the thickness of the barriers and for barriers more than 1 or 2 nanometers thick, it essentially falls to zero inside the barrier (the wall). For barriers thin enough, however, the wave function continues outside the box and there is a finite probability that the electron will show up outside the box. It does not bore a hole through the box, it just shows up outside the box.

---

[6] The wave is a complex number, carrying magnitude and phase information. The probability of the wave being in a given position is $(A^*)A$, where A is the amplitude of the wave and A^* is its complex conjugate.

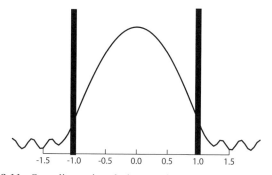

Figure 18.11. One-dimensional electron in a finite box wave function.

Classical (Newtonian) physics cannot explain this phenomenon. The wave function and related probability describe something that has survived innumerable experimental validations. One of the first successes of quantum theory to predict something is the Fowler–Nordheim equation. This equation predicts how many electrons (i.e., what electric current) will flow from a metal surface into a vacuum as a function of the height of the potential barrier at the surface of the metal and the strength of an applied electric field. Fowler and Nordheim showed that an applied electric field distorted the shape of the potential barrier and made it thinner. They correctly predicted the electric current level as a function of the applied field.

Electron tunneling is used today in scanning electron microscopes. A small, sharp, metal tip is used as the source of the electron stream that illuminates the object being studied. Electron tunneling in semiconductors is used in *flash memories*. These are the computer memories that are used in the small "USB dongles" that we plug into our computers to transfer files of pictures or whatever. In a flash memory a voltage is used to narrow a potential barrier so that electrons can transfer onto an isolated electrode. When this voltage is removed, the potential barrier widens again, the tunneling probability drops very, very low, and the electrons are trapped on this electrode—hence the memory effect. There are of course millions or billions of these isolated electrodes on one small chip, giving us our "Mbyte" and "Gbyte" memories. Associated circuitry senses the electron level on trapped electrodes and "reads" the memory without disturbing its content.

The last thing I'd like to present on the subject of quantum randomness is a perspective on how the randomness of fundamental physical processes turns out to have a very deep philosophical impact on how we view our lives and the universe.

Everything going on in the universe is due to particle interactions. Every physical entity is ultimately made up of atoms which in turn are made up of subatomic particles. These particles interact with each other and with photons of electromagnetic radiation (light, x-rays, gamma rays), and everything that happens around us (including us) is the result of these interactions.

Isaac Newton described a universe in which particle interactions were governed by simple mechanical rules: $F = ma$, and so on. About a century after Newton, the French scientist/mathematician Pierre-Simon Laplace pointed out that if we knew the current position and velocity of every particle in the universe, we could calculate the future as well as the past position and velocity of every particle in the universe. Whether or not it is possible to know the position and velocity of every particle in the universe is at best a philosophical puzzle in itself so there really wasn't much philosophical consequence to Laplace's conjecture at that time.

About 100 years later (which is about 100 years ago today) Albert Einstein demonstrated, in his *special theory of relativity*, that no particle could move faster than the speed of light. Since light itself of course moves at the speed of light and light is just electromagnetic radiation in a certain range of wavelengths, all electromagnetic radiation moves at the speed of light.

The speed of light in a vacuum (where it moves fastest) is $300,000,000 = 3 \times 10^8$ meters per second. There are, very approximately, $100,000 = 1 \times 10^5$ seconds in a day. Therefore, light in a vacuum travels $3 \times 10^8 * 1 \times 10^5 = 3 \times 10^{13}$ meters or about 20 billion miles in a day. Compared to this distance, the earth is just a small dot.

Imagine drawing a huge sphere of radius about 20 billion miles. Nothing that's not inside this sphere now, be they particles or radiation, can reach us within a day from now. The number of particles inside this sphere, while still unbelievably enormous, is but a miniscule subset of the number of particles in the universe. While we still aren't capable of doing any detailed calculations on this number of particles, we at least have a countable number of particles to think about.

Going back to Laplace's assertion, but limiting ourselves to the particles that we could interact with within a day, we reach the fantastic conclusion that today's events, and by extension all of our future, are predetermined. There are too many particles for us to calculate anything specific about; but, in principle, since we can know all the current positions and velocities, we can calculate everything that's going to happen.

I myself am reluctant to consider that the future is predetermined. The future contains, along with everything else, my personal thoughts and actions. Could these really all be predetermined?

The assertion of quantum mechanicists that fundamental particle interactions are governed by random probabilities lays Laplace's assertion to rest. If we can't even say when an atom of radium is going to emit a particle and decay, then how can we possibly look at everything around us and know what tomorrow will bring?

When looked at from this perspective, it's quite plausible that truly random events run the universe.

INDEX

2438